Computer-Aided Analysis, Modeling, and Design of Microwave Networks

The Wave Approach

For a complete listing of the *Artech House Microwave Library*,
turn to the back of this book.

Computer-Aided Analysis, Modeling, and Design of Microwave Networks

The Wave Approach

Janusz Dobrowolski

Artech House
Boston • London

Library of Congress Cataloguing-in-Publication Data
Dobrowolski, Janusz.
Computer-aided analysis, modeling, and design of microwave networks: the wave approach/Janusz Dobrowolski.
p. cm.
Includes bibliographical references and index.
ISBN 0-89006-669-8 (alk. paper)
1. Microwave circuits—Computer simulation. 2. Computer-aided design. 3. Electric circuit analysis—Computer programs. 4. JADMIC2. I . Title.
TK7876.D6297 1996 95-49607
621.381'32'028553—dc20 CIP

British Library Cataloguing in Publication Data
Dobrowolski, Januz A.
Computer-aided analysis, modeling, and design of microwave networks: the wave approach
1. Microwave circuits—Design—Data processing
I. Title
621.3'813'0285

ISBN 0-89006-669-8

685 Canton Street
Norwood, MA 02062

International Standard Book Number: 0-89006-669-8
Library of Congress Catalog Number: 95-49607

10 9 8 7 6 5 4 3 2 1

Contents

Preface

This book presents the theoretical foundation as well as practical programming implementation of the wave approach to microwave network analysis and design. The book is divided into two major parts: computer-oriented microwave network theory, which includes nine chapters; and a description and user's manual of JADMIC2, which is a general-purpose computer program for microwave network analysis and design contained in the appendix. The book presents the theory and techniques that have been used in JADMIC2 development.

This publication presents a microwave circuit analysis and design program with open source code. This approach allows the user to know what methods and algorithms have been used in various subroutines. Moreover, all of the circuit element models available in the program are accessible. If necessary they may be modified and made more accurate. Additional elements may be added by inserting additional subroutines. Due to the modular nature of the program code, any changes in element models or analysis algorithms are easy to introduce.

The program may be used very successfully by engineers, researchers, and mostly by students. The open source (FORTRAN and C) increases the educational value of the program by eliminating the feeling of mysteriousness associated with most of the canned commercial packages.

JADMIC2 is a general-purpose program that may be used for the analysis and design of linear microwave circuits in frequency domain. It is used to analyze circuits having topology and elements that are defined by the user. The connection scattering matrix is the basis for the formulation of the network equations in JADMIC2. The circuit element library includes a wide variety of lumped and distributed elements. The allowable one-port elements are two-terminal elements whose impedances or admittances are specified, two-terminal series RLC or parallel RLC networks, short-circuited or open-circuited sections of transmission lines, and one-ports described by their scattering parameter specified at each frequency of the analysis. The biggest variety of JADMIC2 elements are two-ports. These elements are two-ports whose scattering matrix elements are specified at each frequency of the analysis, two-ports whose two-terminal impedances or admittances are specified, series RLC and parallel RLC networks connected in series or in parallel between both ports of the elements, short-circuited and open-circuited parallel stubs, short-circuited and open-circuited series stubs, and sections of transmission lines. JADMIC2 allows for the specification of voltage-controlled and current-controlled current sources and of voltage-controlled and current-controlled voltage sources. The program allows for the specification of any one-port and two-port elements whose scattering parameters are specified at each frequency of the analysis. Thus small-signal scattering parameters for diodes, transistors, and other devices may be read into JADMIC2. Parameters of coupled transmission lines considered as a four-port element may be specified for the program.

To perform noise analysis, active elements of the analyzed circuit have to be specified by frequency-dependent noise parameters.

JADMIC2 performs AC analysis, noise analysis, and optimization of the overall scattering parameters and of the noise figure. The program provides options for optimization of certain network functions or combinations of network functions dependent on circuit element parameters. JADMIC2 handles any circuit topology with distributed, lumped, passive, and active elements.

It is believed that in its present version JADMIC2 can be used very effectively in solving many practical microwave circuit design problems.

The book is intended for graduate courses or as a self-study material for engineers and researchers who are involved in computer-aided design of microwave networks.

Some parts of the text of this book have been used for several years at the Warsaw University of Technology to teach student courses in electrical engineering.

The text of the entire book and the FORTRAN procedures of JADMIC2 program were written and prepared by Janusz A. Dobrowolski. The C procedures of the program were written and linked to FORTRAN code program by Wojciech Ostrowski.

The authors wish to thank their colleagues, especially Dr. Zbigniew Nosal and Mr. Miroslaw Adamski of the Warsaw University of Technology, for valuable discussions and helpful comments.

Janusz A. Dobrowolski
Wojciech Ostrowski
Warsaw University of Technology
January 1996

CHAPTER 1

Introduction

The significance of *computer-aided design* (CAD) techniques in microwave network design is obvious. The advance in *monolithic microwave circuit* (MMIC) technology in the last decade would not be possible without sophisticated CAD procedures and programs. In CAD procedures for microwave circuits the initial design is simulated and optimized. This eliminates time-consuming and costly experimental investigations, which is of significant importance in the design and manufacturing of modern microwave circuits because of the very limited ability to incorporate any modifications in the circuit fabricated by MIC (hybrid) technology. In the case of monolithic microwave circuits their fabrication does not allow any provisions for post-production tuning or adjustment.

The use of computer-aided design software for microwave analysis and design is wide and well established. There are many commercially available microwave CAD packages that are used in industry and at universities for research and educational purposes. Some of them are very sophisticated but also very costly. Most of this software comes in the form of canned computer-aided design packages.

This book presents in detail the theoretical foundation and practical programming implementation for the wave approach to microwave network analysis and design.

In Chapter 2 we introduce wave variables and the scattering matrix. The physical representation of these quantities is presented, too. In addition, we discuss the change of reference impedance in scattering matrix definition, relations between the scattering matrix and other matrix representations of networks, two-port-to-three-port scattering matrix transformations, and transformation formulas for scattering matrices of embedded multiport networks. This theory is applied in computer-aided analysis and modeling procedures of microwave networks.

A computer-oriented microwave circuit analysis method based on the connection scattering matrix is discussed in Chapter 3. The analysis involves evaluation of the scattering parameters of individual elements of the circuit in connection with the information on circuit topology and setting them up in the form of the connection scattering matrix. Microwave circuits with any arbitrary topology may be analyzed using this matrix formalism. Computational algorithms of circuit functions are also presented in detail. We show how a computer program evaluates scattering parameters related to the input and the output ports of the overall circuit, the impedance and the admittance matrices related to the input and the output ports of the overall circuit, the input and the output port reflection coefficients, power gain, transducer power gain and available power gain, insertion loss, voltage transfer gain, and group delay. Special attention is devoted to the stability considerations of microwave circuits described by the scattering matrix. According to this theory it is possible to compute the stability coefficient and the determinant of the overall scattering matrix of a network. If the analyzed circuit is unconditionally stable, one may compute the maximum power gain, that is, the simultaneous conjugate match at the input and the output ports of the overall circuit. In the case of potentially unstable microwave circuits it is necessary to compute parameters that define the stability circles on the generator port reflection coefficient plane as well as on the load reflection coefficient plane. When a circuit is potentially unstable, maximum power gain does not exist and one may compute parameters that define constant power gain and constant available power gain circles.

The noise analysis method of microwave circuits used in the JADMIC2 program is considered in Chapter 4. The method is based on the noise wave representation for noise generated in the network. Noise properties of a circuit are described by the noise wave correlation matrices of circuit elements. Noise wave correlation matrices for common elements used for microwave network modeling are also presented. In addition, we discuss the chain matrix noise representation, impedance and admittance matrix representations, and their mutual interrelations. We state relations that describe two-port-to-three-port noise wave transformations and transformation formulas for noise wave correlation matrices of embedded multiport networks. These relations are essential in noise modeling problems of microwave semiconductor devices, such as MESFETs, HEMTs, and BJTs using the wave approach. A detailed discussion of MESFET and HEMT noise modeling using noise waves is given. In Chapter 4 we also discuss algorithms for computing the noise figure and four noise parameters F_{min}, R_n, and Γ_{opt} for multiport microwave networks. The theory related to the constant noise figure circles are also given. In the last section of this chapter we present relations for constant noise measure circle computations.

In Chapter 5 we present the theory related to the transposed matrix method for sensitivity analysis of microwave networks described by the connection scattering matrix. Our discussion concentrates on the computational algorithm of the scattering parameter sensitivities with respect to circuit parameters such as characteristic impedances and lengths of transmission lines, resistances, inductances, and capacitances. In addition, derivations of sensitivity expressions for network elements commonly used in microwave circuits are presented.

The method for evaluating group delay of scattering matrix transfer parameters based on the transposed matrix principle is given. In the last two sections of this chapter we present the method for noise power sensitivity analysis of microwave networks, which may be used for noise figure gradient computation. The noise figure gradient vector is applicable in gradient optimization methods and in computer-aided tolerance analysis and design of microwave networks.

Chapter 6 is devoted to a short description of sparse matrix techniques used in the JADMIC2 program for the very effective computation of repeated solutions of circuit equations related to frequency domain analysis, sensitivity analysis, and noise analysis of a microwave circuit described by the connection scattering matrix. The JADMIC2 sparse matrix techniques are based on the reduced bifactorization method. The JADMIC2 practical implementation of the method uses a looping indexed code technique.

In Chapter 7 we present the basics of optimization techniques used in CAD of microwave networks and discuss the basics for formulating the generalized least *p*th objective function, which is very convenient in defining design objectives of microwave networks in frequency and time domains. Because unconstrained optimization will often produce unacceptable parameter values that have no physical or practical meaning, the realistic design of a network must account for constraints imposed on design network parameters. These problems may be solved using the transformation of optimized variables.

Chapter 8 addresses time domain analysis methods of linear microwave networks. The chapter contains short descriptions of Fourier series and the Fourier transform mathematical basis and a full description of time domain network analysis based on the connection scattering matrix and the wave approach.

Chapter 9 presents a theory of balance methods used to solve the practical problem of finding a periodic steady-state response of microwave nonlinear network excited by a source of a periodic "large" signal of a given frequency ω_0 and DC sources biasing nonlinear devices (transistors and diodes). The multireflection method for solving the harmonic balance equation utilizing an incident wave as a parametrizing variable is discussed in detail.

The last chapter of the book discusses noise wave parameter measurement methods for active microwave two-ports. We present concepts of noise parameter analyzers based on an assumption of direct measurement of noise wave parameters (noise wave correlation matrices) of a device under test. Thanks to this assumption all difficulties and drawbacks related to standard noise parameter measurement methods based on the application of a source-pull tuner may be overcome. We present two such analyzers: a six-port noise parameter analyzer and an interferometer system for noise wave parameter measurements. Both systems are particularly suitable for on-wafer and on-chip measurements of microwave active device two-port parameters.

The Appendixes are the user's manual of JADMIC2, which is a general-purpose computer program for microwave network analysis and design. The attached diskette contains open source code of the program. This approach allows the user to know what methods and algorithms have been used in various subroutines. In addition, all circuit element models available in the program are accessible. If necessary they may be modified and made more accurate. Additional elements may be added by inserting additional subroutines. Due to the modular nature of the program code, any changes in element models or analysis algorithms can be easily introduced.

JADMIC2 was developed out of a motivation for maximum efficiency. The basis for the program is the connection scattering matrix, the sparse matrix techniques, and the transposed matrix method approach for computing network sensitivities and a gradient of a given performance index.

JADMIC2 performs AC analysis, noise analysis, and optimization of the overall scattering parameters and the noise figure. The program provides options for optimization of certain network functions or combinations of network functions dependent on circuit element parameters. JADMIC2 handles any circuit topology with distributed, lumped, passive, and active elements.

To operate JADMIC2, the user must create and store a data file that presents his circuit to the computer by:

1. Defining the type of each circuit element;
2. Describing every circuit element with its parameter numerical data;
3. Describing the circuit topology, that is, interconnections between circuit elements;
4. Defining the frequency range of the analysis/design;
5. Selecting the type of desired outputs;
6. In the case of optimization, specifying the target results.

JADMIC2 is very simple to learn and use. A typical microwave circuit designer can learn to operate JADMIC2 within a few hours without any programming background.

In designing the interface with the user, several facts have been taken into account. First, JADMIC2 is an interactive program. Therefore, the user may modify, change, and refine his design in a concise manner and still be able to find the solution to the design problem. The user is able, for example, to change the circuit very easily, run as many analyses as desired, and change optimization goals.

Second, the designer has the options to define and describe the circuit in a meaningful and simple fashion. JADMIC2 uses easy-to-remember shorthand mnemonics (key words) by means of which the user defines, for example, types of circuit elements, values of element parameters, circuit topology, frequency range of interest, and optimization goals.

Third, the user of JADMIC2 has the complete presentation of output results in tabular form as well as in graph form. The tabular form is exact, while the graph form gives immediate insight. Therefore, video display is used for immediate results and hard copy plots for subsequent analysis.

It is believed that in its present version JADMIC2 can be used very effectively in solving many practical microwave circuit design problems.

CHAPTER 2

Wave-Based Representation of Microwave Networks

A microwave network is a set of multiports connected to each other. The components (elements) of a microwave network may be one-, two-, three-, or four-ports, in general, multiports. In the wave variable approach, network characterization is accomplished using scattering matrices. As it will be shown later, the scattering matrix formulation is very convenient for the description and for the computer-aided analysis and design of microwave circuits with any topology.

The scattering matrix description for microwave circuits will be reviewed in this chapter. Relationships for transformations among admittance, impedance, and scattering matrix representations will be presented. We provide a table that will contain scattering matrices for some basic and commonly encountered elements of microwave networks.

2.1 SCATTERING MATRIX REPRESENTATION

In 1965 K. Kurokawa defined complex power wave variables and introduced generalized scattering parameters [1]. The scattering parameters for real and positive reference impedances of multiport ports were defined earlier by P. Penfield [2, 3] and D. Youla [4].

The scattering parameters describe the interrelationship of a new set of variables called power waves (a_i, b_i). The ingoing and outgoing power waves at the ith port of the network are defined in terms of the terminal voltage V_i, the terminal current I_i, and the arbitrary reference impedance Z_{Ri} ($\mathrm{Re}\{Z_{Ri}\} > 0$), as [1]

$$a_i = \frac{V_i + Z_{Ri} I_i}{2\sqrt{\text{Re}\{Z_{Ri}\}}} \tag{2.1}$$

$$b_i = \frac{V_i - Z_{Ri}^* I_i}{2\sqrt{\text{Re}\{Z_{Ri}\}}} \tag{2.2}$$

where the asterisk denotes the complex conjugate.

The inverse transformations to (2.1) and (2.2) are of the forms

$$V_i = \frac{1}{\sqrt{\text{Re}\{Z_{Ri}\}}} (Z_{Ri}^* a_i + Z_{Ri} b_i) \tag{2.3}$$

$$I_i = \frac{1}{\sqrt{\text{Re}\{Z_{Ri}\}}} (a_i - b_i) \tag{2.4}$$

which means that for a fixed Z_{Ri}, if V_i and I_i are given, the power wave variables a_i and b_i can be calculated from (2.1) and (2.2); and vice versa, the power waves a_i and b_i can easily be converted into V_i and I_i using (2.3) and (2.4).

As a result of (2.1) and (2.2) we define the power wave reflection coefficient:

$$\Gamma = \frac{b_i}{a_i} = \frac{V_i - Z_{Ri}^* I_i}{V_i + Z_{Ri} I_i} = \frac{V_i/I_i - Z_{Ri}^*}{V_i/I_i + Z_{Ri}} \tag{2.5}$$

Because $V_i/I_i = Z_L$ (see Figure 2.1),

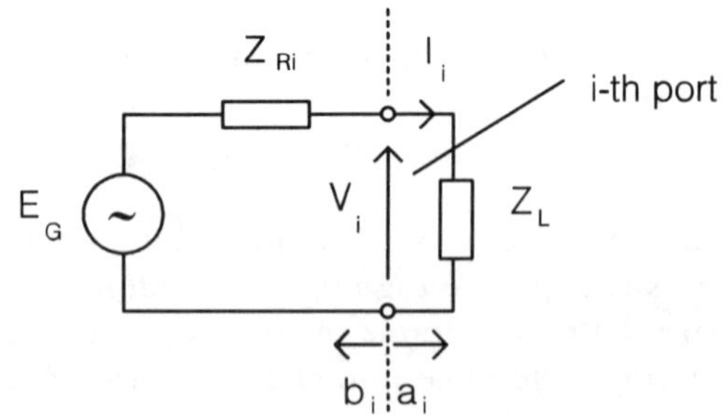

Figure 2.1 A circuit illustrating the relationship between power waves and terminal voltage and current at a port.

$$\Gamma = \frac{Z_L - Z_{Ri}^*}{Z_L + Z_{Ri}} \tag{2.6}$$

where Z_L is the load impedance (input impedance of the *i*th port) and Z_{Ri} is the reference impedance of the *i*th port.

The physical interpretation of complex wave variables a_i and b_i comes from the following discussion.

Referring to Figure 2.1, we have

$$V_i = E_G - Z_{Ri} I_i \tag{2.7}$$

Inserting this into (2.1), we obtain

$$|a_i|^2 = \frac{|E_G|^2}{4\,\mathrm{Re}\{Z_{Ri}\}} = P_{GA} \tag{2.8}$$

Next let us consider a quantity $|a_i|^2 - |b_i|^2$. Direct substitution of (2.1) and (2.2) into this expression gives

$$\begin{aligned} |a_i|^2 - |b_i|^2 &= \frac{1}{4\,\mathrm{Re}\{Z_{Ri}\}}[(V_i + Z_{Ri} I_i)(V_i^* + Z_{Ri}^* I_i^*) - (V_i - Z_{Ri}^* I_i)(V_i^* - Z_{Ri} I_i^*)] \\ &= \frac{1}{4\,\mathrm{Re}\{Z_R\}}\{Z_{Ri} + Z_{Ri}^*\}(V_i I_i^* + V_i^* I_i) \\ &= \mathrm{Re}\{V_i I_i^*\} = P_L = P_{GA} - P_R \end{aligned} \tag{2.9}$$

The quantity $|a_i|^2 - |b_i|^2$ can also be written in the form

$$|a_i|^2 - |b_i|^2 = |a_i|^2\left(1 - \left|\frac{b_i}{a_i}\right|^2\right) = |a_i|^2(1 - |\Gamma|^2) \tag{2.10}$$

It follows from (2.8) to (2.10) that the value of $|a_i|^2$ at the port of the load impedance Z_L is equal to the available power of the signal generator with the internal impedance equal to the port reference impedance Z_{Ri}.

The absolute value $|a_i|^2$ is the power that the signal generator sends toward a load, regardless of the load impedance. Equation (2.9) shows that the real power dissipated in the load impedance Z_L equals $|a_i|^2 - |b_i|^2$. Then, (2.8) to (2.10) can be interpreted as follows. The signal generator sends the power $|a_i|^2$ toward the load, regardless of the load impedance. When the load is not matched (i.e., $\Gamma \neq 0$, (2.7)), a part of the incident power is reflected back to the generator. In the case when the load is matched (i.e., when $Z_L = Z_{Ri}^*$), the total power sent by the generator dissipates in the load.

Under the matching condition:

$$Z_L = Z_{Ri}^* \tag{2.11}$$

we have $b_i = 0$, $\Gamma = 0$, and $P_L = |a_i|^2 = P_{GA}$.

Referring to Figure 2.2 the power wave scattering matrix of a multiport with n ports is defined by the equation

$$\mathbf{b} = \mathbf{S}\,\mathbf{a} \tag{2.12}$$

where **S** is a square matrix of order n and **a** and **b** are vectors, respectively, of ingoing and outgoing power wave variables at ports of the multiport.

In the case when the reference impedances of the circuit ports Z_{Ri}, $i = 1, 2, \ldots, n$, are real and equal to the characteristic impedances Z_{oi} of the transmission lines connected to the multiport ports, the definitions of a_i and b_i simplify to the definitions of complex ingoing and outgoing voltage (or current) waves in a transmission line and the power wave scattering matrix **S** becomes the microwave scattering matrix.

Substituting

$$V_i = V_i^+ + V_i^- \tag{2.13}$$

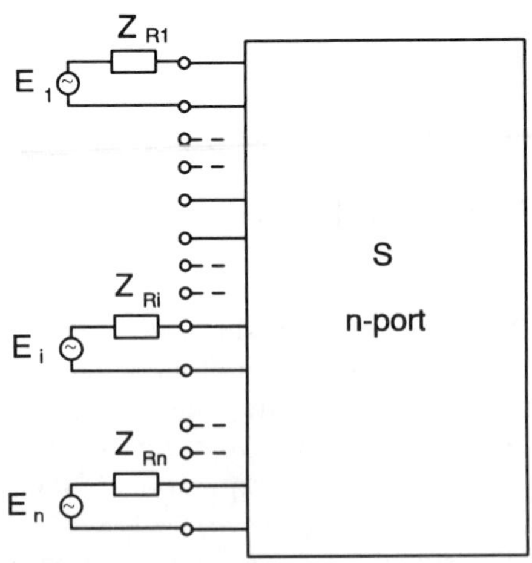

Figure 2.2 A multiport excited at its ports by voltage sources with internal impedances equal to the reference impedances of these ports.

$$I_i = I_i^+ + I_i^- \tag{2.14}$$

$$Z_{Ri} = Z_{oi} = \frac{V_i^+}{I_i^+} = \frac{V_i^-}{I_i^-} \tag{2.15}$$

to (2.1) and (2.2), we obtain

$$a_i = \frac{V_i^+}{\sqrt{Z_{oi}}} = \sqrt{Z_{oi}} I_i^+ \tag{2.16}$$

$$b_i = \frac{V_i^-}{\sqrt{Z_{oi}}} = \sqrt{Z_{oi}} I_i^- \tag{2.17}$$

$$\Gamma = \frac{b_i}{a_i} = \frac{Z_L - Z_{oi}}{Z_L + Z_{oi}} \tag{2.18}$$

In the above relations V_i^+ and I_i^+ denote the ingoing voltage and current waves, respectively, whereas V_i^- and I_i^- are the outgoing voltage and current waves at the *i*th port of the network. Microwave scattering parameters can be measured using well-known methods and measurement systems [5].

All conveniences coming from the assumption that the reference impedance Z_{Ri} is real and positive are usually used in computer programs for microwave network analysis and design. It is convenient that all variables and scattering parameters of a network are referenced to a single positive real impedance Z_R common to all network ports.

2.2 PHYSICAL INTERPRETATION OF SCATTERING PARAMETERS

Let P_i denote the real power delivered to the *i*th port of a multiport. According to (2.9), this power is given as [6–8]

$$P_i = |a_i|^2 - |b_i|^2 \tag{2.19}$$

The total power delivered to all n ports of the circuit describes the relation

$$P = \sum_{i=1}^{n} P_i = \sum_{i=1}^{n} |a_i|^2 - \sum_{i=1}^{n} |b_i|^2 = \mathbf{a}^+\mathbf{a} - \mathbf{a}^+\mathbf{a} \tag{2.20}$$

where the superscript $^+$ indicates the complex conjugate transposed matrix (vector).

Substituting $\mathbf{b} = \mathbf{S}\ \mathbf{a}$ into (2.20), we have

$$P = \mathbf{a}^{+} \left[\mathbf{I} - \mathbf{S}^{+}\mathbf{S} \right] \mathbf{a} \tag{2.21}$$

For a passive multiport, in which $P \geq 0$, the scattering matrix **S** must satisfy the condition

$$\mathbf{I} - \mathbf{S}^{+}\mathbf{S} \geq \mathbf{0} \tag{2.22}$$

for all real ω. In the case of a passive and lossless multiport, $P = 0$ and

$$\mathbf{I} - \mathbf{S}^{+}\mathbf{S} = \mathbf{0} \tag{2.23}$$

Therefore, the scattering matrix of a passive and lossless multiport is a unitary matrix (in (2.21) to (2.23) **I** indicates a unit matrix).

According to the matrix equation (2.12), the elements of the scattering matrix **S** can be derived from the relations

$$S_{ii} = \left.\frac{b_i}{a_i}\right|_{a_k = 0,\ k \neq i} \tag{2.24}$$

and

$$S_{ki} = \left.\frac{b_k}{a_i}\right|_{a_k = 0,\ k \neq i} \tag{2.25}$$

The condition $a_k = 0$, $k \neq i$, means that each circuit port, except port i, is not excited by a signal generator and simultaneously is terminated in its reference impedance, Z_{Ri}, so that the waves b_k outgoing from these ports are totally absorbed in these terminations. Figure 2.3 presents such a circuit. Its kth port is excited by the voltage source E_k with the internal impedance equal to the reference impedance of this port.

Except the excited port, for all other ports $V_k = -I_k Z_{Rk}$, and according to (2.1) to (2.2)

$$a_k = 0 \tag{2.26}$$

and

$$b_k = -\sqrt{Z_{Rk}}\, I_k = \frac{1}{\sqrt{Z_{Rk}}} V_k \tag{2.27}$$

For the excited port, $E_i = V_i + Z_{Ri} I_i$ and, according to (2.1), the ingoing wave is

$$a_i = \frac{E_i}{2\sqrt{Z_{Ri}}} \tag{2.28}$$

In order to find the physical interpretation of the S_{ii} scattering parameter, let Z_{INi} be the input impedance at the *i*th port of a network whose all other ports are terminated with their reference impedances (Figure 2.3). Because $V_i = Z_{INi} I_i$, according to (2.1) we have

$$a_i = \frac{(Z_{INi} + Z_{Ri})I_i}{2\sqrt{Z_{Ri}}} \tag{2.29}$$

and

$$b_i = \frac{(Z_{INi} - Z_{Ri})I_i}{2\sqrt{Z_{Ri}}} \tag{2.30}$$

In this way we have

$$S_{ii} = \left.\frac{b_i}{a_i}\right|_{a_k = 0,\ k \neq i} = \frac{Z_{INi} - Z_{Ri}}{Z_{INi} + Z_{Ri}} \tag{2.31}$$

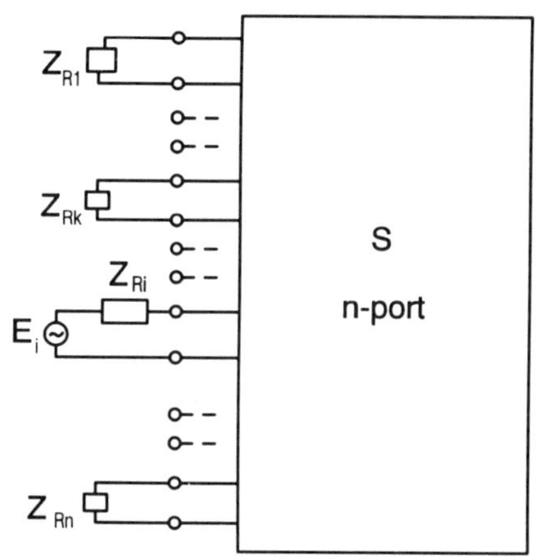

Figure 2.3 Multiport excited at the *i*th port with other ports terminated in its reference impedances Z_{Ri}.

In words, S_{ii} is the power wave reflection coefficient at the ith port of the multiport, whose other ports are terminated with their reference impedances.

The squared value of the magnitude of S_{ii} equals the power reflection coefficient

$$|S_{ii}|^2 \left|\frac{b_i}{a_i}\right|^2\Bigg|_{a_k=0,\ k\neq i} = \frac{P_{\mathrm{R}i}}{P_{\mathrm{GA}i}}\Bigg|_{a_k=0,\ k\neq i} \tag{2.32}$$

where $P_{\mathrm{GA}i}$ is the available power of the signal generator with the internal impedance $Z_{\mathrm{R}i}$ connected to the ith port and $P_{\mathrm{R}i}$ is the power reflected from ith port when all other ports are matched.

Now let us consider the physical interpretation of the S_{ki} scattering parameter. Using (2.27) and (2.28) in (2.25), we have

$$S_{ki} = \frac{b_k}{a_i}\Bigg|_{a_k=0,\ k\neq i} = -2\sqrt{Z_{\mathrm{R}i}Z_{\mathrm{R}k}}\,\frac{I_k}{E_i} \tag{2.33}$$

When the reference impedances of the ith and kth ports are real and equal to each other ($Z_{\mathrm{R}i} = Z_{\mathrm{R}k} = Z_{\mathrm{R}}$), the transmission coefficient S_{ki} equals the voltage gain $2V_k/E_i$ between the ports i and k when all circuit ports are terminated at their reference impedances.

It is usually assumed that reference impedances of all element ports of analyzed network are real and equal to each other. It means that S_{ki} equals the voltage gain

$$S_{ki} = 2\,\frac{V_k}{E_i} \tag{2.34}$$

Setting the magnitude of S_{ki} to the second power we get

$$|S_{ki}|^2 = \left|\frac{b_k}{a_i}\right|^2\Bigg|_{a_k=0,\ k\neq i} = \frac{Z_k|I_k|^2}{|E_i|^2/(4\ Z_i)} = \frac{P_{\mathrm{L}k}}{P_{\mathrm{GA}i}} = G_{\mathrm{T}} \tag{2.35}$$

where $P_{\mathrm{GA}i}$ is the available power of the signal generator connected to the ith port and $P_{\mathrm{L}k}$ is the real power dissipated in the load $Z_{\mathrm{L}k} = Z_{\mathrm{R}k}$ of the kth port of the circuit.

In accordance with the above relations, $|S_{ki}|^2$ equals the transducer power gain G_{T} between the kth port and the ith port of a multiport all of whose ports are loaded with impedances equal to the reference impedances of these ports.

Group delay is a circuit function of great importance for microwave filter and equalizer characterization.

For transmission from port i to port k, the group delay is defined as

$$\tau_{ki} = -\frac{\partial \Phi_{ki}}{\partial \omega} \tag{2.36}$$

where Φ_{ki} is the phase of S_{ki}.

The relation for τ_{ki} may be expressed as

$$\tau_{ki} = -\mathrm{Im}\left[\frac{1}{S_{ki}}\frac{\partial S_{ki}}{\partial \omega}\right] \tag{2.37}$$

Some of the most important properties of the scattering matrix are as follows.

1. The **S** matrix of a reciprocal network is symmetrical; that is,

$$\mathbf{S} = \mathbf{S}^{\mathrm{T}} \tag{2.38}$$

where T denotes the operation of the transposition of a matrix. The condition (2.38) restricts the number of independent elements of the **S** matrix of the order n to $(n^2 + n)/2$.

2. The **S** matrix of a lossless network is the unitary matrix

$$\mathbf{S}^{+}\mathbf{S} = \mathbf{I} \tag{2.39}$$

Example 2.1

Let us consider a circuit presented in Figure 2.4 and find its scattering parameters using definitions of wave variables and the scattering matrix. We assume that reference impedances of both ports are real and equal to each other.

The S_{11} parameter may be found using the relation

$$S_{11} = \left.\frac{b_1}{a_1}\right|_{a_2=0}$$

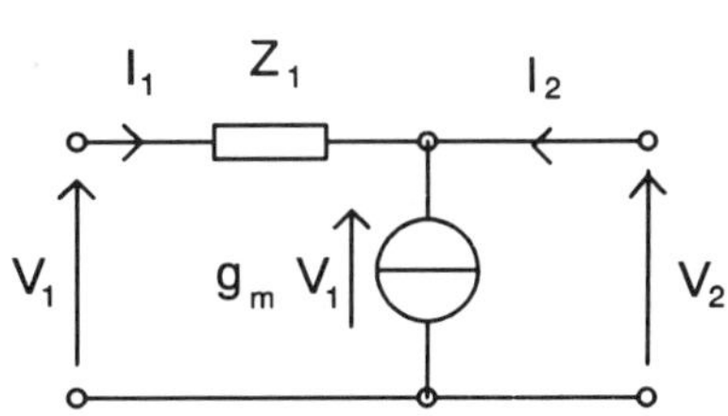

Figure 2.4 A circuit.

Because a_2 must be equal to zero, the output port of the circuit must be terminated with the load impedance equal to the reference impedance Z_R.

Using Figure 2.5 and (2.1) and (2.2) we have

$$S_{11} = \frac{V_1 - Z_R I_1}{V_1 + Z_R I_1} = \frac{V_1/I_1 - Z_R}{V_1/I_1 + Z_R} = \frac{Z + g_m Z_R^2}{Z + 2Z_R - g_m Z_R^2}$$

The S_{21} parameter is defined by the relation

$$S_{21} = \left.\frac{b_2}{a_1}\right|_{a_2 = 0}$$

Using Figure 2.5 and (2.1) and (2.2) we get

$$S_{21} = \frac{V_2 - Z_R I_2}{V_1 + Z_R I_1} = \frac{2Z_R(1 - g_m Z)}{Z + Z_R(2 - g_m Z_R)}$$

The next two parameters are S_{12} and S_{22}. Because

$$S_{12} = \left.\frac{b_1}{a_2}\right|_{a_1 = 0}$$

using relations shown in Figure 2.6 we have

$$S_{12} = \frac{V_1 - Z_R I_1}{V_2 + Z_R I_2} = \frac{2Z_R}{2Z_R + Z(1 + g_m Z_R)}$$

Similarly

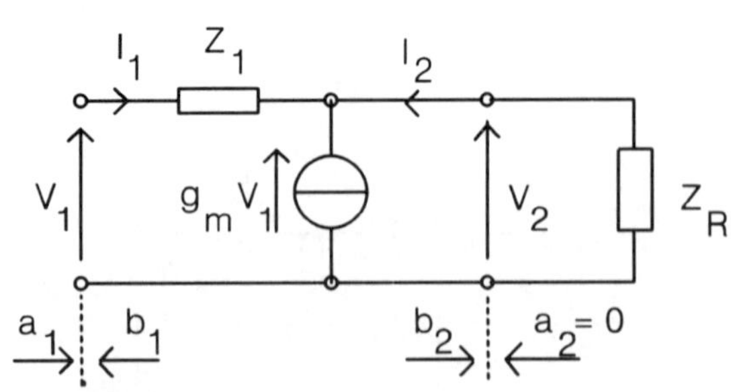

Figure 2.5 A circuit for determining S_{11} and S_{21}. The circuit is excited at the input port and terminated with the reference impedance at the output port.

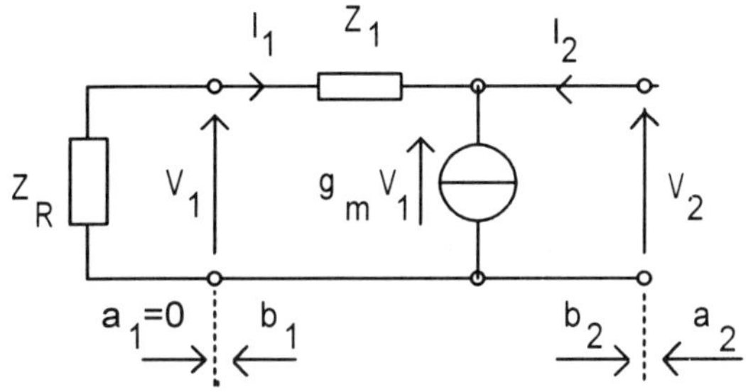

Figure 2.6 A circuit for determining S_{22} and S_{12}. The circuit is excited at the output port and terminated with the reference impedance at the input port.

$$S_{22} = \left.\frac{b_2}{a_2}\right|_{a_1 = 0}$$

and using Figure 2.6 we have

$$S_{22} = \frac{V_2 - Z_R I_2}{V_2 + Z_R I_2} = \frac{Z + Z_R - Z_R(1 - g_m Z_R)}{Z + Z_R + Z_R(1 - g_m Z_R)}$$

2.3 CHANGE OF PORT REFERENCE IMPEDANCE IN THE SCATTERING MATRIX REPRESENTATION OF A MULTIPORT

Suppose that the port reference impedances of a multiport under consideration are changed from Z_{Ri} to Z'_{Ri}, $i = 1, 2, \ldots, n$. Then according to (2.1) and (2.2) the ingoing and outgoing waves must be redefined accordingly. Of course, the scattering matrix $\mathbf{S}'$ relating these new power wave variables is different from the original one. It is expressed in terms of the original $\mathbf{S}$ matrix as [1]

$$\mathbf{S}' = \mathbf{A}^{-1}\ (\mathbf{S} - \Gamma^{+})\ (\mathbf{I} - \Gamma\ \mathbf{S})^{-1}\mathbf{A}^{+} \tag{2.40}$$

where Γ is a diagonal matrix whose ith diagonal element equals

$$r_i = \frac{Z'_{Ri} - Z_{Ri}}{Z'_{Ri} - Z^{*}_{Ri}} \tag{2.41}$$

A is also a diagonal matrix. It can be shown that the ith diagonal element A_i is

$$A_i = \frac{1 - r_i^*}{|1 - r_i|}\sqrt{|1 - r_i r_i^*|} \tag{2.42}$$

Formula (2.40) finds practical applications. For example, in the design of two-port transistor amplifier whose source and load impedances are Z_{R1} and Z_{R2}, respectively, the matching conditions for the input and output ports are given by $S'_{11} = 0$ and $S'_{22} = 0$. Using (2.40) the condition $S'_{11} = 0$ gives

$$r_1^* = \frac{S_{11} - r_2(S_{11}S_{22} - S_{12}S_{21})}{1 - r_2 S_{22}} \tag{2.43}$$

Similarly, $S'_{22} = 0$ provides the relation

$$r_2^* = \frac{S_{22} - r_1(S_{11}S_{22} - S_{12}S_{21})}{1 - r_1 S_{11}} \tag{2.44}$$

For the simultaneous match at the input and the output of a two-port network, (2.43) and (2.44) must be satisfied simultaneously.

The scattering matrices for some basic elements of microwave circuits are presented in Table 2.1.

2.4 RELATIONS BETWEEN DIFFERENT MATRIX REPRESENTATIONS OF MULTIPORTS

2.4.1 Relationships Between the S Matrix and the Y and Z Matrices

Let us refer to the circuit shown in Figure 2.2 and assume that internal impedances of all signal generators connected to the network ports are equal to the reference impedances of individual ports. According to the definition of complex wave variables ((2.1) and (2.2)), vectors of ingoing power waves **a** and outgoing power waves **b** at the circuit ports can be written in the forms [1]

$$\mathbf{a} = \mathbf{F}\,(\mathbf{v} + \mathbf{G}\,\mathbf{i}) \tag{2.45}$$

$$\mathbf{b} = \mathbf{F}\,(\mathbf{v} - \mathbf{G}^{+}\mathbf{i}) \tag{2.46}$$

In the above equations **v** and **i** are vectors whose ith elements are equal to the terminal voltages V_i, $i = 1, 2, \ldots, n$, and to the terminal currents I_i, $i = 1, 2, \ldots, n$,

Table 2.1
Scattering Matrices of Some Circuit Elements

Circuit element	Scattering matrix S
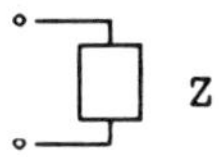	$\dfrac{Z - Z_R}{Z + Z_R}$
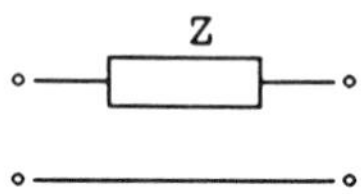	$\dfrac{1}{Z + 2\,Z_R}\begin{bmatrix} Z & 2\,Z_R \\ 2\,Z_R & Z \end{bmatrix}$
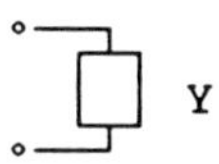	$\dfrac{1 - Z_R^{*}\,Y}{1 + Z_R\,Y}$
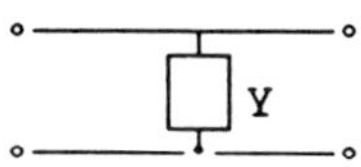	$\dfrac{1}{Y + 2\,Y_R}\begin{bmatrix} -Y & 2\,Y_R \\ 2\,Y_R & -Y \end{bmatrix}$
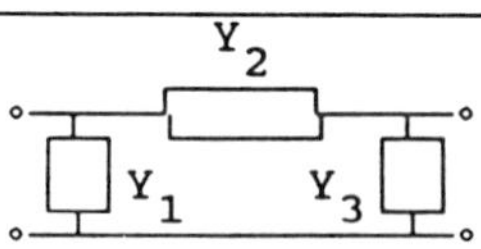 Π - network	$\dfrac{1}{D}\begin{bmatrix} Y_R^2 - PY_R - T & 2Y_RY_3 \\ 2Y_RY_3 & Y_R^2 + PY_R - T \end{bmatrix}$ $D = Y_R^2 + QY_R + T$ $\quad P = Y_1 - Y_2$ $T = Y_1Y_2 + Y_2Y_3 + Y_1Y_3$ $\quad Q = Y_1 + Y_2 + 2Y_3$
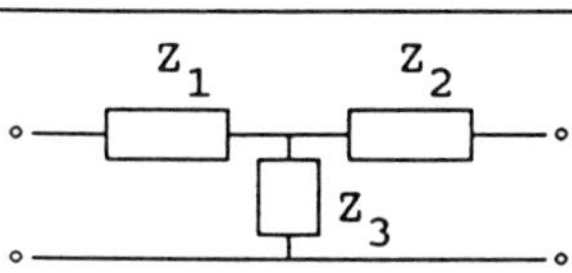 T - network	$\dfrac{1}{D}\begin{bmatrix} -Z_R^2 + PZ_R + T & 2Z_RZ_3 \\ 2Z_RZ_3 & -Z_R^2 - PZ_R + T \end{bmatrix}$ $D = Z_R^2 + QZ_R + T$ $\quad P = Z_1 - Z_2$ $T = Z_1Z_2 + Z_2Z_3 + Z_1Z_3$ $\quad Q = Z_1 + Z_2 + 2Z_3$
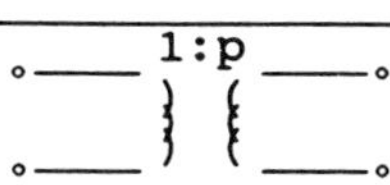 Ideal transformer	$\dfrac{1}{1 + p^2}\begin{bmatrix} 1 - p^2 & 2\,p \\ 2\,p & p^2 - 1 \end{bmatrix}$
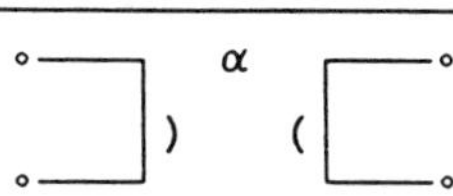 Gyrator	$\dfrac{1}{\alpha^2 + Z_R^2}\begin{bmatrix} \alpha^2 - Z_R^2 & 2\,\alpha\,Z_R \\ -2\,\alpha\,Z_R & \alpha^2 - Z_R^2 \end{bmatrix}$

Table 2.1 (continued)

Circuit element	Scattering matrix S
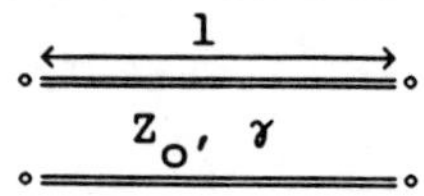 Transmission line section $\gamma = \alpha + j\beta \quad , \beta = \frac{2\Pi}{\lambda_g}$	$\frac{1}{Z_o^2 + Z_R^2 + 2 Z_o Z_R \coth(\gamma l)} \cdot \begin{bmatrix} Z_o^2 - Z_R^2 & 2 Z_o Z_R \operatorname{csch}(\gamma l) \\ 2 Z_o Z_R \operatorname{csch}(\gamma l) & Z_o^2 - Z_R^2 \end{bmatrix}$
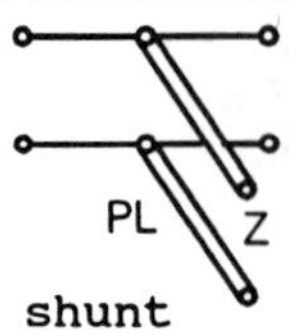 Open ended shunt connected stub	$\frac{1}{2 Y_R + Y_o \tanh(\gamma l)} \cdot \begin{bmatrix} -Y_o \tanh(\gamma l) & 2 Y_R \\ 2 Y_R & -Y_o \tanh(\gamma l) \end{bmatrix}$
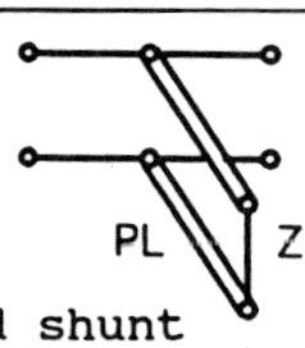 Short ended shunt connected stub	$\frac{1}{2 Y_R + Y_o \coth(\gamma l)} \cdot \begin{bmatrix} -Y_o \coth(\gamma l) & 2 Y_R \\ 2 Y_R & -Y_o \coth(\gamma l) \end{bmatrix}$
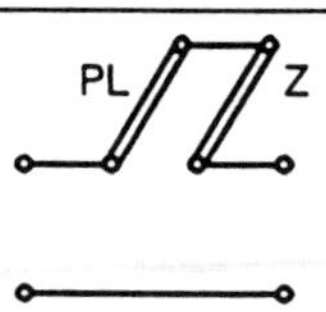 Short ended series connected stub	$\frac{1}{2Z_R + Z_0 \tanh(\gamma l)} \cdot \begin{bmatrix} Z_0 \tanh(\gamma l) & 2Z_R \\ 2Z_R & Z_0 \tanh(\gamma l) \end{bmatrix}$
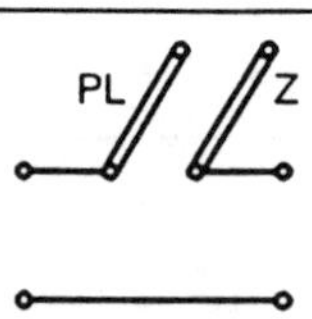 Open ended series connected stub	$\frac{1}{2Z_R + Z_0 \coth(\gamma l)} \cdot \begin{bmatrix} Z_0 \coth(\gamma l) & 2Z_R \\ 2Z_R & Z_0 \coth(\gamma l) \end{bmatrix}$

Table 2.1 (continued)

Circuit element	Scattering matrix S
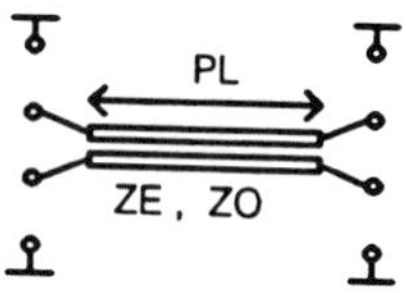 Z_{oe}, Z_{oo}, l $\gamma_e = \alpha_e + j\beta_e,\ \beta_e = \frac{2\Pi}{\lambda_{ge}}$ $\gamma_o = \alpha_o + j\beta_o,\ \beta_o = \frac{2\Pi}{\lambda_{go}}$ A section of coupled transmission lines	$\begin{bmatrix} S_{11} & S_{12} & S_{13} & S_{14} \\ S_{12} & S_{11} & S_{14} & S_{13} \\ S_{13} & S_{14} & S_{11} & S_{12} \\ S_{14} & S_{13} & S_{12} & S_{11} \end{bmatrix}$ $S_{11} = \frac{S_{11e} + S_{11o}}{2} \quad S_{12} = \frac{S_{11e} - S_{11o}}{2}$ $S_{13} = \frac{S_{12e} - S_{12o}}{2} \quad S_{14} = \frac{S_{12e} + S_{12o}}{2}$ $S_{11e,o} = \frac{Z_{oe,o}^2 - Z_R^2}{Z_{oe,o}^2 + Z_R^2 + 2Z_{oe,o}Z_R\coth(\gamma_{e,o}l)}$ $S_{12e,o} = \frac{2Z_{oe,o}\ Z_R\ \mathrm{csch}(\gamma_{e,o}l)}{Z_{oe,o}^2 + Z_R^2 + 2Z_{oe,o}Z_R\coth(\gamma_{e,o}l)}$
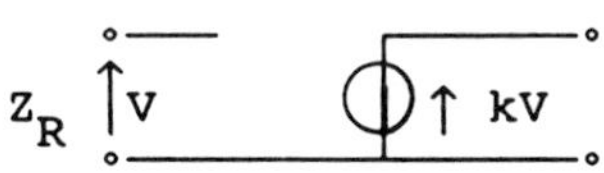 Voltage controlled voltage source	$\begin{bmatrix} 1 & 0 \\ 2k & -1 \end{bmatrix}$
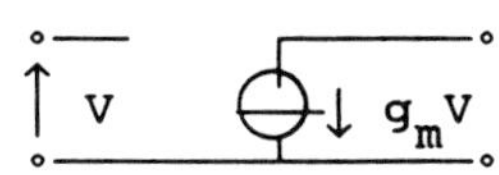 Voltage controlled current source	$\begin{bmatrix} 1 & 0 \\ -2Z_R\ g_m & 1 \end{bmatrix}$
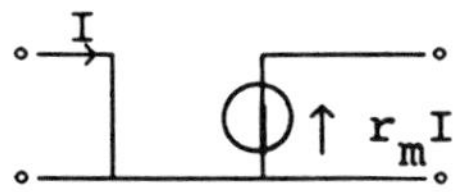 Current controlled voltage source	$\begin{bmatrix} -1 & 0 \\ \frac{2r_m}{Z_R} & -1 \end{bmatrix}$
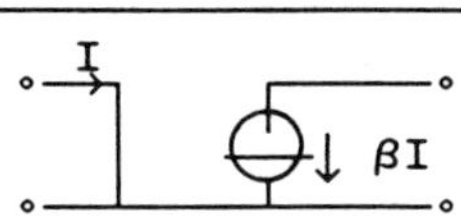 Current controlled current source	$\begin{bmatrix} -1 & 0 \\ -2\beta & 1 \end{bmatrix}$

Table 2.1 (continued)

Circuit element	Scattering matrix S
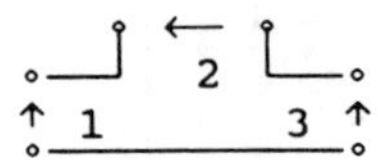 Series connection of three ports	$\frac{1}{3}\begin{bmatrix} 1 & 2 & 2 \\ 2 & 1 & -2 \\ 2 & -2 & 1 \end{bmatrix}$
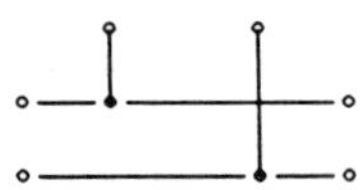 Parallel conection of three ports	$\frac{1}{3}\begin{bmatrix} -1 & 2 & 2 \\ 2 & -1 & 2 \\ 2 & 2 & -1 \end{bmatrix}$
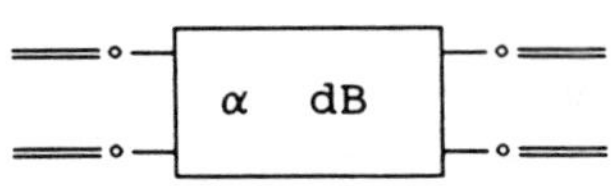 Attenuator α dB	$\begin{bmatrix} 0 & A \\ A & 0 \end{bmatrix}$ $A = 10^{-\alpha/20}$
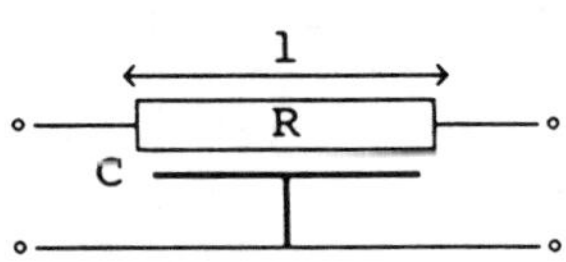 Distributed RC line section	As for a transmission line section with $Z_0 = \sqrt{\frac{R}{j\omega C}}$ and $\gamma l = \sqrt{j\omega RC}$ $R = r\cdot l$ - total resistance, $C = c\cdot l$ - total capacitance.

respectively, at the ports of the n-port circuit. **F** and **G** are the diagonal matrices given by

$$F=\begin{bmatrix} \frac{1}{2\sqrt{\mathrm{Re}\ Z_{R1}}} & 0 & \cdots & 0 \\ 0 & \frac{1}{2\sqrt{\mathrm{Re}\ Z_{R2}}} & & \vdots \\ \vdots & & \ddots & 0 \\ 0 & \cdots & & \frac{1}{2\sqrt{\mathrm{Re}\ Z_{Rn}}} \end{bmatrix} \tag{2.47}$$

and

$$G=\begin{bmatrix} Z_{R1} & 0 & \cdots & 0 \\ 0 & Z_{R2} & & \vdots \\ \vdots & & \ddots & \vdots \\ \vdots & & & 0 \\ 0 & \cdots & 0 & Z_{Rn} \end{bmatrix} \tag{2.48}$$

Because $\mathbf{b} = \mathbf{S}\,\mathbf{a}$, $\mathbf{v} = \mathbf{Z}\,\mathbf{i}$, and $\mathbf{i} = \mathbf{Y}\,\mathbf{v}$, using relations (2.45) and (2.46) we obtain

$$\mathbf{F}\,(\mathbf{Z} - \mathbf{G}^{+})\,\mathbf{i} = \mathbf{S}\,\mathbf{F}\,(\mathbf{Z} + \mathbf{G})\,\mathbf{i} \tag{2.49}$$

from which we have the relation between **S** and **Z**

$$\mathbf{S} = \mathbf{F}\,(\mathbf{Z} - \mathbf{G}^{+})\,(\mathbf{Z} + \mathbf{G})^{-1}\mathbf{F}^{-1} \tag{2.50}$$

Similarly, the impedance matrix **Z** can be expressed in terms of the scattering matrix **S** as

$$\mathbf{Z} = \mathbf{F}^{-1}\,(\mathbf{I} - \mathbf{S})^{-1}(\mathbf{S}\,\mathbf{G} + \mathbf{G}^{+})\,\mathbf{F} \tag{2.51}$$

The relations between the **S** and **Y** matrices are found in the same way. We have

$$\mathbf{S} = \mathbf{F}\,(\mathbf{I} - \mathbf{G}^{+}\mathbf{Y})\,(\mathbf{I} + \mathbf{G}\,\mathbf{Y})^{-1}\mathbf{F}^{-1} \tag{2.52}$$

and

$$\mathbf{Y} = \mathbf{F}(\mathbf{I} - \mathbf{S})\,(\mathbf{S}\,\mathbf{G} + \mathbf{G}^{+})^{-1}\mathbf{F}^{-1} \tag{2.53}$$

In (2.51) to (2.53) **I** denotes the identity matrix.

The above relations are used for computing the overall impedance matrix and the overall admittance matrix from the overall scattering matrix of the analyzed circuit.

2.4.2 Relationships Between the S Matrix and the T (*ABCD*) Matrix

The relations describing the transformation between the **S** and **T** (*ABCD*) matrices [9] are

$$A = \frac{Z_{R1}^* + Z_{R1}S_{11} - Z_{R1}^* S_{22} - Z_{R1}\Delta_S}{2\, S_{21}\sqrt{\mathrm{Re}\{Z_{R1}\}\,\mathrm{Re}\{Z_{R2}\}}} \tag{2.54}$$

$$B = \frac{Z_{R1}^* + Z_{R2}^* + Z_{R1}Z_{R2}^* S_{11} + Z_{R1}^* Z_{R2}S_{22} + Z_{R1}Z_{R2}\Delta_S}{2\, S_{21}\sqrt{\mathrm{Re}\{Z_{R1}\}\,\mathrm{Re}\{Z_{R2}\}}} \tag{2.55}$$

$$C = \frac{1 - S_{11} - S_{22} + \Delta_S}{2\, S_{21}\sqrt{\mathrm{Re}\{Z_{R1}\}\,\mathrm{Re}\{Z_{R2}\}}} \tag{2.56}$$

$$D = \frac{Z_{R2}^* - Z_{R2}^* S_{11} + Z_{R2}S_{22} - Z_{R2}\Delta_S}{2\, S_{21}\sqrt{\mathrm{Re}\{Z_{R1}\}\,\mathrm{Re}\{Z_{R2}\}}} \tag{2.57}$$

where Z_{R1} and Z_{R2} are the reference impedances of two-port ports and $\Delta_s = S_{11}S_{22} - S_{12}S_{21}$.

The inverse relations are

$$S_{11} = \frac{A\, Z_{R2} + B - C\, Z_{R1}^* Z_{R2} - D\, Z_{R1}^*}{A\, Z_{R2} + B + C\, Z_{R1}Z_{R2} + D\, Z_{R1}} \tag{2.58}$$

$$S_{12} = \frac{2\,(AD - BC)\,\sqrt{\mathrm{Re}\{Z_{R1}\}\,\mathrm{Re}\{Z_{R2}\}}}{A\, Z_{R2} + B + C\, Z_{R1}Z_{R2} + D\, Z_{R1}} \tag{2.59}$$

$$S_{21} = \frac{2\,\sqrt{\mathrm{Re}\{Z_{R1}\}\,\mathrm{Re}\{Z_{R2}\}}}{A\, Z_{R2} + B + C\, Z_{R1}Z_{R2} + D\, Z_{R1}} \tag{2.60}$$

$$S_{22} = \frac{-A\, Z_{R2}^* + B - C\, Z_{R1}Z_{R2}^* + D\, Z_{R1}}{A\, Z_{R2} + B + C\, Z_{R1}Z_{R2} + D\, Z_{R1}} \tag{2.61}$$

2.5 TWO-PORT-TO-THREE-PORT SCATTERING MATRIX TRANSFORMATION

The two-port S parameters of a three-terminal device, like BJT or MESFET, with one terminal grounded can be converted to three-port scattering parameters when the grounded terminal is used as a third terminal [10].

As shown in Figure 2.7, incident and reflected power waves are referred by matrix equation in the form

$$\begin{bmatrix} b_1 \\ b_2 \\ b_3 \end{bmatrix} = \begin{bmatrix} S_{11} & S_{12} & S_{13} \\ S_{21} & S_{22} & S_{23} \\ S_{31} & S_{32} & S_{33} \end{bmatrix} \begin{bmatrix} a_1 \\ a_2 \\ a_3 \end{bmatrix} \tag{2.62}$$

Assuming that port 2 is constrained by the reflection coefficient $\Gamma_2 = -1$ ($\Gamma_2 = -1$ means grounding the terminal 2),

$$a_2 = -b_2 \tag{2.63}$$

Substituting (2.63) into (2.62) and then eliminating b_2 yields

$$b_1 = \left[S_{11} - \frac{S_{13}\,S_{21}}{1 + S_{33}} \right] a_1 + \left[S_{12} - \frac{S_{13}\,S_{32}}{1 + S_{33}} \right] a_2 \tag{2.64}$$

$$b_2 = \left[S_{21} - \frac{S_{23}\,S_{31}}{1 + S_{33}} \right] a_1 + \left[S_{22} - \frac{S_{23}\,S_{32}}{1 + S_{33}} \right] a_2 \tag{2.65}$$

From (2.64) and (2.65) it is now possible to obtain expressions for standard two-port parameters

$$S'_{11} = S_{11} - \frac{S_{13}\,S_{21}}{1 + S_{33}} \tag{2.66}$$

$$S'_{12} = S_{12} - \frac{S_{13}\,S_{32}}{1 + S_{33}} \tag{2.67}$$

$$S'_{21} = S_{21} - \frac{S_{23}\,S_{31}}{1 + S_{33}} \tag{2.68}$$

$$S'_{22} = S_{22} - \frac{S_{23}\,S_{32}}{1 + S_{33}} \tag{2.69}$$

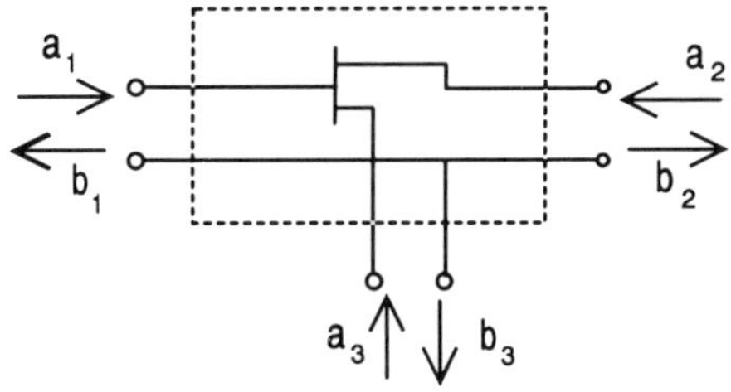

Figure 2.7 Incident and reflected waves at three ports of a three-terminal device treated as a three-port.

It is important to mention here that the nine elements of coefficient matrix **S** in (2.62) are not independent if we consider a three-terminal device as a three-port. In fact, there are only four independent parameters. If these are known, the others can be found because they are all related by the condition [10]

$$\sum_{i=1}^{3} S_{ij} = \sum_{j=1}^{3} S_{ij} = 1 \tag{2.70}$$

This theory indicates how a set of the two-port scattering parameters in any common configuration (common source, common drain, or common gate) can be converted into any other set of the two-port parameters in another common configuration.

2.6 SCATTERING MATRICES OF AN EMBEDDED MULTIPORT NETWORK

In active device modeling problems it is very common to consider separately an equivalent circuit of the intrinsic device and an equivalent circuit of the extrinsic chip. The equivalent circuit of an intrinsic device contains mostly controlled voltage or controlled current sources responsible for device amplification, while the equivalent circuit of the extrinsic chip contains mostly parasitic elements such as stay capacitances and lead inductances and resistances. The effect of additional elements on the scattering matrix of the entire device may be calculated using the following embedding formulas. Figure 2.8 presents two subnetworks connected by internal ports.

The scattering matrix for the aggregate network is found by partitioning the embedding network's scattering matrix **T** into submatrices

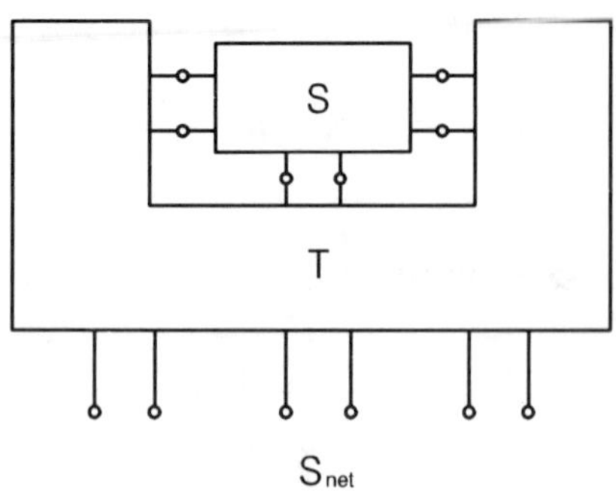

Figure 2.8 A multiport network with scattering matrix **S** embedded within the network with scattering matrix **T**. $\mathbf{S}_{net}$ is the scattering matrix of the entire network referred to external ports.

$$\begin{bmatrix} \mathbf{b}_e \\ \mathbf{b}_i \end{bmatrix} = \begin{bmatrix} \mathbf{T}_{ee} & \mathbf{T}_{ei} \\ \mathbf{T}_{ie} & \mathbf{T}_{ii} \end{bmatrix} \begin{bmatrix} \mathbf{a}_e \\ \mathbf{a}_i \end{bmatrix} \tag{2.71}$$

where $\mathbf{a}_e$, $\mathbf{b}_e$ and $\mathbf{a}_i$, $\mathbf{b}_i$ are vectors of incoming and outgoing waves, respectively, in the external and internal ports of the network.

The embedded subnetwork describes the system of equations

$$\mathbf{b}_s = \mathbf{S}\ \mathbf{a}_s \tag{2.72}$$

The connections of pairs of internal ports of **S** and **T** subnetworks impose restrictions on vectors of internal incoming and outgoing waves of the forms

$$\mathbf{a}_i = \mathbf{b}_s \tag{2.73}$$

$$\mathbf{b}_i = \mathbf{a}_s \tag{2.74}$$

From (2.71), (2.72), (2.73), and (2.74), by first eliminating $\mathbf{b}_i$, we obtain

$$\mathbf{a}_i = (\mathbf{S}^{-1} - \mathbf{T}_{ii})^{-1}\ \mathbf{T}_{ie}\ \mathbf{a}_e \tag{2.75}$$

and, after eliminating $\mathbf{a}_i$ from the first equation of (2.71)

$$\mathbf{b}_e = [\mathbf{T}_{ee} + \mathbf{T}_{ei}\ (\mathbf{S}^{-1} - \mathbf{T}_{ii})^{-1}\ \mathbf{T}_{ie}]\ \mathbf{a}_e \tag{2.76}$$

The coefficient matrix in this equation

$$\mathbf{S}_{net} = \mathbf{T}_{ee} + \mathbf{T}_{ei}\ (\mathbf{S}^{-1} - \mathbf{T}_{ii})^{-1}\ \mathbf{T}_{ie} \tag{2.77}$$

is the scattering matrix of the whole network referred to the external ports.

It is convenient to write (2.77) in the form

$$\mathbf{S}_{net} = \mathbf{T}_{ee} + \Lambda\ \mathbf{S}\ \mathbf{T}_{ie} \tag{2.78}$$

where Λ is the matrix given by

$$\Lambda = \mathbf{T}_{ei}\ (\mathbf{I} - \mathbf{S}\ \mathbf{T}_{ii})^{-1} \tag{2.79}$$

In (2.79) **I** is the identity matrix.

During the computation of the resultant scattering matrix $\mathbf{S}_{net}$, the majority of computation time is spent on matrix inversion required for Λ matrix evaluation.

References

[1] Kurokawa, K., "Power Waves and the Scattering Matrix," *IEEE Trans. Microwave Theory Tech.*, Vol. MTT-13, No. 2, March 1965, pp. 194–202.

[2] Penfield, P., "Noise in Negative Resistance Amplifiers," *IRE Trans. Circuit Theory*, Vol. CT-7, 1960, pp. 166–170.

[3] Penfield, P., "A Classification of Lossless Three Ports," *IRE Trans. Circuit Theory*, Vol. CT-9, 1963, pp. 215–223.

[4] Youla, D. C., "On Scattering Matrices Normalized to Complex Port Numbers," *Proc. IRE*, Vol. 49, 1961, p. 122.

[5] Hewlett-Packard, Application Note 117–1, 1969.

[6] Montgomery, C. G., R. H. Dicke, and E. M. Purcell, *Principles of Microwave Circuits*, New York: McGraw-Hill, 1948.

[7] Altman, J. L., *Microwave Circuits*, Princeton, NJ: Van Nostrand, 1964.

[8] Ramo, S., J. R. Whinnery, and T. Van Duzer, *Fields and Waves in Communication Electronics*, 2nd ed., New York: Wiley, 1984.

[9] Dobrowolski, J. A., *Introduction to Computer Methods for Microwave Circuit Analysis and Design*, Norwood, MA: Artech House, 1991.

[10] Bodway, G. E., "Circuit Design and Characterization of Transistors by Means of Three-Port Scattering Parameters," *Microwave J.*, May 1968, pp. 55–63.

CHAPTER 3

Frequency Domain Analysis of Microwave Networks Using Wave Computation

By the analysis of a microwave network we understand the computation of a certain number of response functions in terms of the element parameters for the network with fixed topology. Network equations are formulated by means of normalized wave variables. Microwave networks may contain lumped elements as well as distributed elements. The equivalent circuit models of active devices are mostly origins of lumped components in microwave networks. Obviously, lumped resistances, capacitances, and inductances can be present in these circuits [1, 2]. All circuit elements are considered as multiports characterized in terms of their scattering matrices. Since a microwave circuit may contain lumped as well as distributed elements, it is convenient for computer-aided analysis to characterize both types of elements as multiport equivalents. It is very well known that multiterminal lumped elements can be transformed exactly into multiport circuits [1].

The objective of microwave network analysis is to determine reflection coefficients at certain ports as well as transmission coefficients between selected ports, usually the input port and the load port. The connection scattering matrix method is a very convenient approach to CAD microwave network analysis. At each frequency of interest, scattering matrices are formed for each of the network elements. The circuit topology then dictates how these matrices form the overall circuit connection scattering matrix **W**. Using this last matrix as a coefficient matrix of the system of linear equations, it is possible to obtain ingoing and outgoing waves at all network ports.

3.1 CONNECTION SCATTERING MATRIX METHOD

As it is shown in Figure 3.1, a microwave network is an interconnection of m elements (multiports).

For the kth circuit element we have a set of linear equations

$$\mathbf{b}^{(k)} = \mathbf{S}^{(k)} \mathbf{a}^{(k)} \tag{3.1}$$

where $\mathbf{S}^{(k)}$ is its scattering matrix and $\mathbf{a}^{(k)}$ and $\mathbf{b}^{(k)}$ are vectors of ingoing and outgoing wave variables, respectively, at its ports.

An independent signal generator is described by the equation

$$b_G = S_G a_G + c \tag{3.2}$$

and is illustrated in Figure 3.2.

Substituting into the equation

$$V_G = E_G + I_G Z_G \tag{3.3}$$

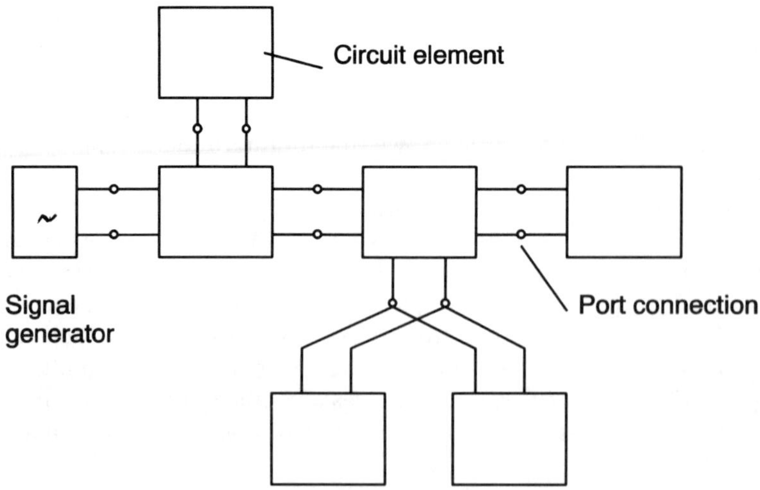

Figure 3.1 Microwave network containing interconnected multiports and independent signal generators.

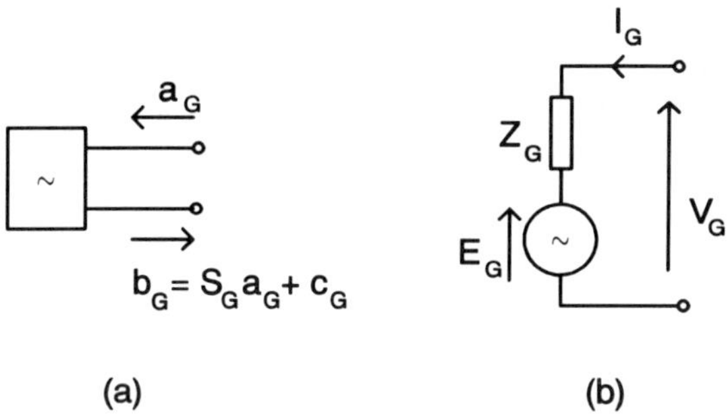

Figure 3.2 Independent signal generator: (a) incoming and outgoing waves at the generator port, and (b) equivalent sinusoidal signal voltage source.

the relations

$$V_G = \frac{1}{\sqrt{Z_R}}(Z_R\, a_G + Z_R\, b_G)$$

$$I_G = \frac{1}{\sqrt{Z_R}}(a_G - b_G) \tag{3.4}$$

which relate the voltage V_G across and the current I_G flowing into the independent generator port with the incoming wave a_G and outgoing wave b_G at that port, we get

$$b_G = \frac{Z_G - Z_R}{Z_G + Z_R}\, a_G + \frac{\sqrt{Z_R}\, E_G}{Z_G + Z_R} \tag{3.5}$$

where Z_R is the reference impedance of the generator port.

Comparing (3.2) and (3.5) we see that

$$S_G = \frac{Z_G - Z_R}{Z_G + Z_R} \tag{3.6}$$

is the reflection coefficient of the independent generator port and

$$c = \frac{\sqrt{Z_R}\, E_G}{Z_G + Z_R} \tag{3.7}$$

is the independent complex wave impressed by the generator.

The magnitude of c set to the second power equals the power transferred from the generator to the load, whose impedance equals the reference impedance Z_R of the generator port.

Considering a whole microwave network with m elements, we have a set of linear equations in the matrix form [3–6]

$$\mathbf{S}\,\mathbf{a} + \mathbf{c} = \mathbf{b} \tag{3.8}$$

where

$$\mathbf{a} = \begin{bmatrix} \mathbf{a}^{(1)} \\ \mathbf{a}^{(2)} \\ \vdots \\ \mathbf{a}^{(k)} \\ \vdots \\ \mathbf{a}^{(m)} \end{bmatrix} \qquad \mathbf{b} = \begin{bmatrix} \mathbf{b}^{(1)} \\ \mathbf{b}^{(2)} \\ \vdots \\ \mathbf{b}^{(k)} \\ \vdots \\ \mathbf{b}^{(m)} \end{bmatrix} \qquad \mathbf{c} = \begin{bmatrix} \mathbf{c}^{(1)} \\ \mathbf{c}^{(2)} \\ \vdots \\ \mathbf{c}^{(k)} \\ \vdots \\ \mathbf{c}^{(m)} \end{bmatrix} \tag{3.9}$$

and

$$\mathbf{S} = \begin{bmatrix} \mathbf{S}^{(1)} & \mathbf{O} & \dots & \mathbf{O} & \dots & \mathbf{O} \\ \mathbf{O} & \mathbf{S}^{(2)} & & & & \vdots \\ \vdots & & \ddots & & & \\ \mathbf{O} & \dots & & \mathbf{S}^{(k)} & \dots & \mathbf{O} \\ \vdots & & & & \ddots & \vdots \\ \mathbf{O} & \dots & & \mathbf{O} & \dots & \mathbf{S}^{(m)} \end{bmatrix} \tag{3.10}$$

In (3.10) $\mathbf{S}^{(1)}, \mathbf{S}^{(2)}, \dots, \mathbf{S}^{(m)}$ are the scattering matrices of multiports or the reflection coefficients of independent generators and loads. They describe all m elements of a microwave network. The vectors $\mathbf{a}^{(1)}, \mathbf{a}^{(2)}, \dots, \mathbf{a}^{(m)}$; $\mathbf{b}^{(1)}, \mathbf{b}^{(2)}, \dots, \mathbf{b}^{(m)}$; and $\mathbf{c}^{(1)}, \mathbf{c}^{(2)}, \dots, \mathbf{c}^{(m)}$ represent wave variables related to these elements.

Taking into account the connections between the m individual elements of the network we must introduce constraints on vectors $\mathbf{a}$ and $\mathbf{b}$ that can be represented by the matrix equation

$$\mathbf{b} = \mathbf{\Gamma}\,\mathbf{a} \tag{3.11}$$

where $\mathbf{\Gamma}$ is the connection matrix of the network.

Substituting (3.11) into (3.8) we obtain

$$\mathbf{W}\,\mathbf{a} = \mathbf{c} \tag{3.12}$$

where

$$\mathbf{W} = \mathbf{\Gamma} - \mathbf{S} \tag{3.13}$$

The coefficient matrix **W** in (3.13) is called the connection scattering matrix of the network. The right-hand side vector **c** is the vector of the impressed waves of the independent signal generators in the network. The solution of the matrix equation (3.12) is the vector of incoming waves at all ports of the analyzed network

$$\mathbf{a} = \mathbf{W}^{-1}\,\mathbf{c} \tag{3.14}$$

Substituting (3.14) into (3.11) we can compute a vector of outgoing waves in all ports of the network as

$$\mathbf{b} = \mathbf{\Gamma}\,\mathbf{W}^{-1}\,\mathbf{c} \tag{3.15}$$

The set of linear equations (3.12) with the connection scattering matrix **W** as the coefficient matrix allows us to compute the incoming and outgoing waves at all network ports when the excitations **c** of the network are known. The connection scattering matrix represents the full description of the network because values of the scattering parameters of all the network elements are given in the matrix **S** and information about the network topology is given in the matrix **Γ**.

To find the principles of the formation of the connection matrix **Γ** we consider the simple case of a connection of two ports. It is shown in Figure 3.3.

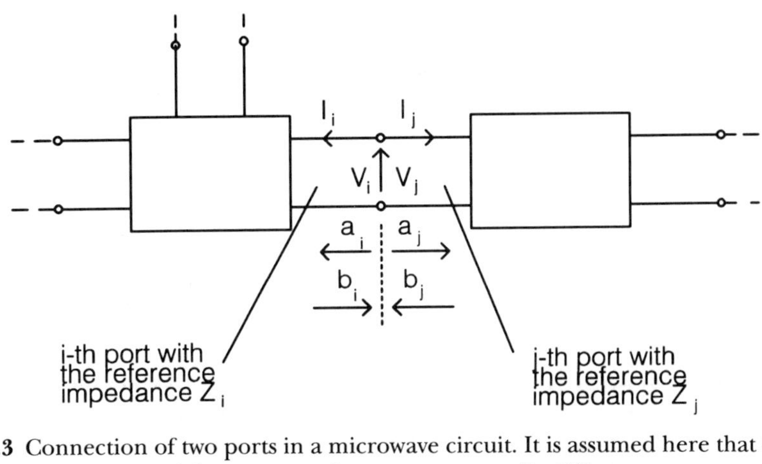

Figure 3.3 Connection of two ports in a microwave circuit. It is assumed here that reference impedances of the connected ports are generally different.

The connection between the ith port and jth port of the circuit imposes the following relations in the plane of the port connection

$$V_i = V_j \quad \text{and} \quad -I_i = I_j \tag{3.16}$$

Substituting into (3.16) the relations for currents and voltages as functions of ingoing and outgoing waves on both sides of the port connection ((3.19) and (3.20)), we get

$$\begin{bmatrix} b_i \\ b_j \end{bmatrix} = \frac{1}{Z_i + Z_j} \begin{bmatrix} Z_j - Z_i & 2\sqrt{Z_i Z_j} \\ 2\sqrt{Z_i Z_j} & Z_i - Z_j \end{bmatrix} \begin{bmatrix} a_i \\ a_j \end{bmatrix} \tag{3.17}$$

The coefficient matrix in (3.17) is the connection matrix $\mathbf{\Gamma}$ for one pair of the connected ports. In the same way we express the connection matrices of each pair of the connected pairs of ports from which, after appropriate ordering, we can formulate the connection matrix $\mathbf{\Gamma}$ of the whole circuit.

From (3.17) we observe that if

$$Z_i = Z_j \tag{3.18}$$

the connection of two ports is a nonreflecting connection, which means that the outgoing and ingoing waves at ports i and j connected together satisfy the relations

$$a_i = b_j \quad \text{and} \quad a_j = b_i \tag{3.19}$$

It follows from these considerations that when the reference impedances of the connected ports satisfy condition (3.18) and, of course, when the reference impedances of the connected ports are real and equal to each other, the connection matrix of such a pair of ports simplifies to the form

$$\mathbf{\Gamma} = \begin{bmatrix} 0 & 1 \\ 1 & 0 \end{bmatrix} \tag{3.20}$$

Also the elements of the connection matrix of the whole network are ones and zeros. As we see, the connection matrix $\mathbf{\Gamma}$ is a sparse matrix in which most of the elements are equal to zero. Only the elements γ_{ij} and γ_{ji} corresponding to pairs of the connected ports are ones.

As was previously mentioned, it is usually assumed that all reference impedances of all network ports are real and equal to each other.

The algorithm of the formation of connection matrix $\mathbf{\Gamma}$ of a microwave circuit is not complicated when the connections between element ports are restricted to the connections of pairs of ports. In the case of connections of three or more ports in one plane, a form of the connection matrix **Γ** and also the algorithm of its formation are much more complicated. Because of this, it is more convenient to limit the number of the connected ports in one plane to two. In all cases, when in the analyzed network there are connections of three or more ports in one plane, it is appropriate to treat such a connection as a separate network element with given scattering matrix **S**. For example, the parallel connection of three ports in one plane (presented in Figure 3.4), treated as a network element, has the scattering matrix given by (when the reference impedances of all ports are real and equal to each other)

$$\mathbf{S} = \frac{1}{3}\begin{bmatrix} -1 & 2 & 2 \\ 2 & -1 & 2 \\ 2 & 2 & -1 \end{bmatrix} \tag{3.21}$$

We treat the series connection of three ports in one plane in a similar way. It is also a network element, and its scattering matrix is equal to (see Figure 3.4(b))

$$\mathbf{S} = \frac{1}{3}\begin{bmatrix} 1 & 2 & 2 \\ 2 & 1 & -2 \\ 2 & -2 & 1 \end{bmatrix} \tag{3.22}$$

It is assumed here that the series connection element is excited at port 1.

The computer-aided analysis method that is based on the connection scattering matrix requires the multiple computation of the solution of a system of linear equations in which the order of the coefficient matrix equals the number of ports of all network elements. Conventional numerical procedures for the solution of sets of linear equations require $n^3/3 + n^2 - n/3$ multiplications and divisions (where n is the order of the coefficient matrix). A considerable reduction in the execution time of multiple repeated numerical solutions of network equations (3.12) for different circuit parameters, various frequencies, or for different excitations of the circuit (different **c** vector) can be achieved using the sparse matrix techniques. Sparse

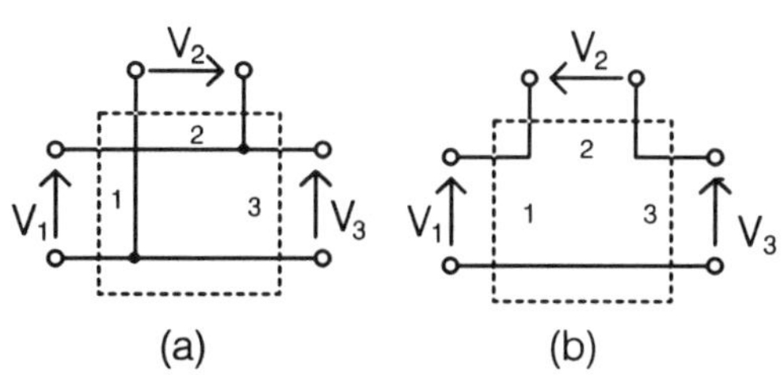

Figure 3.4 Definition of two types of connections of three ports in one plane: (a) parallel connection, and (b) series connection.

matrix techniques also allow a reduction in the memory space required to store the nonzero entries of the coefficient matrix of a system of linear equations to be solved.

Example 3.1

Consider the microwave transistor amplifier in Figure 3.5. The scattering matrix **S** and the connection matrix **Γ** of the network are

$$\mathbf{S} = \begin{array}{c|cccccccc} & 1 & 2 & 3 & 4 & 5 & 6 & 7 & 8 \\ \hline 1 & S_{11} & S_{12} & 0 & 0 & 0 & 0 & 0 & 0 \\ 2 & S_{21} & S_{22} & 0 & 0 & 0 & 0 & 0 & 0 \\ 3 & 0 & 0 & S_{33} & S_{34} & 0 & 0 & 0 & 0 \\ 4 & 0 & 0 & S_{43} & S_{44} & 0 & 0 & 0 & 0 \\ 5 & 0 & 0 & 0 & 0 & S_{55} & S_{56} & 0 & 0 \\ 6 & 0 & 0 & 0 & 0 & S_{65} & S_{66} & 0 & 0 \\ 7 & 0 & 0 & 0 & 0 & 0 & 0 & S_{77} & 0 \\ 8 & 0 & 0 & 0 & 0 & 0 & 0 & 0 & S_{88} \end{array}$$

$$\mathbf{\Gamma} = \begin{array}{c|cccccccc} & 1 & 2 & 3 & 4 & 5 & 6 & 7 & 8 \\ \hline 1 & 0 & 0 & 0 & 1 & 0 & 0 & 0 & 0 \\ 2 & 0 & 0 & 0 & 0 & 1 & 0 & 0 & 0 \\ 3 & 0 & 0 & 0 & 0 & 0 & 0 & 1 & 0 \\ 4 & 1 & 0 & 0 & 0 & 0 & 0 & 0 & 0 \\ 5 & 0 & 1 & 0 & 0 & 0 & 0 & 0 & 0 \\ 6 & 0 & 0 & 0 & 0 & 0 & 0 & 0 & 1 \\ 7 & 0 & 0 & 1 & 0 & 0 & 0 & 0 & 0 \\ 8 & 0 & 0 & 0 & 0 & 0 & 1 & 0 & 0 \end{array}$$

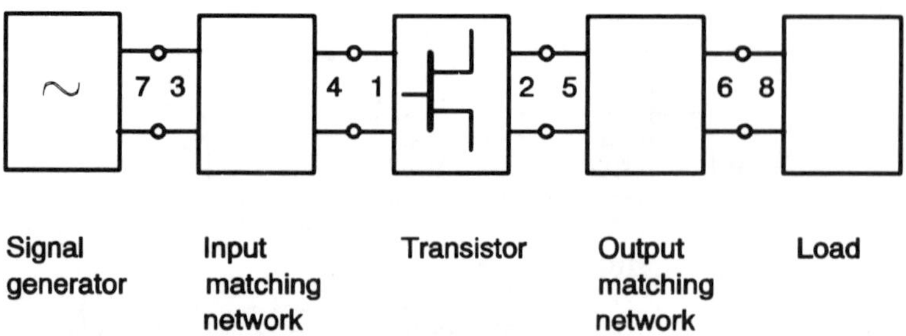

Figure 3.5 Microwave transistor amplifier. Three two-ports and two one-ports connected by their ports create the network.

The connection scattering matrix **W** and the vector of excitations **c** have the following forms. Note that the excitation vector **c** has only one nonzero element, c_7, which corresponds to the generator port number 7.

$$
\mathbf{W} = \begin{array}{c} \\ 1 \\ 2 \\ 3 \\ 4 \\ 5 \\ 6 \\ 7 \\ 8 \end{array}
\begin{array}{c}
\begin{array}{cccccccc} 1 & 2 & 3 & 4 & 5 & 6 & 7 & 8 \end{array} \\
\left[\begin{array}{cccccccc}
-S_{11} & -S_{12} & 0 & 1 & 0 & 0 & 0 & 0 \\
-S_{21} & -S_{22} & 0 & 0 & 1 & 0 & 0 & 0 \\
0 & 0 & -S_{33} & -S_{34} & 0 & 0 & 1 & 0 \\
1 & 0 & -S_{43} & -S_{44} & 0 & 0 & 0 & 0 \\
0 & 2 & 0 & 0 & -S_{55} & -S_{56} & 0 & 0 \\
0 & 0 & 0 & 0 & -S_{65} & -S_{66} & 0 & 1 \\
0 & 0 & 3 & 0 & 0 & 0 & -S_{77} & 0 \\
0 & 0 & 0 & 0 & 0 & 1 & 0 & -S_{88}
\end{array}\right]
\end{array}
\qquad
\mathbf{c} = \begin{array}{c} 1 \\ 2 \\ 3 \\ 4 \\ 5 \\ 6 \\ 7 \\ 8 \end{array}
\left[\begin{array}{c} 0 \\ 0 \\ 0 \\ 0 \\ 0 \\ 0 \\ 1 \\ 0 \end{array}\right]
$$

3.2 COMPUTATION OF NETWORK RESPONSE FUNCTIONS

The analysis method based on the connection scattering matrix allows the determination of incident and reflected waves at all network ports. In terms of these variables all response functions of the network may be derived. Here we consider the basics of computing many network response functions.

3.2.1 Input Port Reflection Coefficient

Let the circuit shown in Figure 3.6 be driven by a signal generator with the impedance Z_G and terminated by the load with the impedance Z_L.

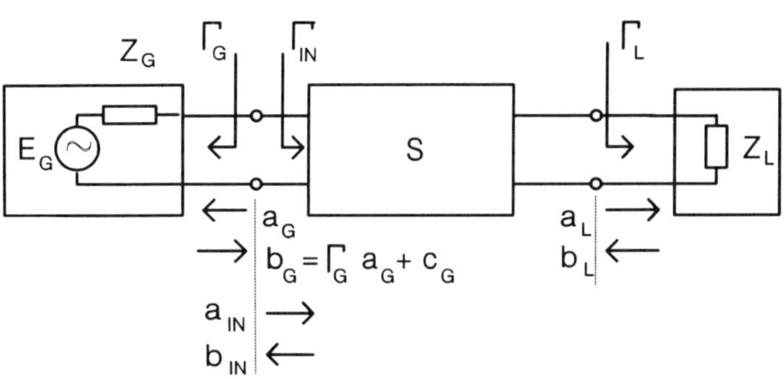

Figure 3.6 A two-port driven by a signal generator and terminated by a load Z_L with the reflection coefficient Γ_L. The network model for computing input port reflection coefficient Γ_{IN}. Γ_{IN} is a function of **S** and Γ_L.

The incident and reflected waves at the generator port are related by the equation

$$b_G = S_G \, a_G + c \tag{3.23}$$

where

$$S_G = \frac{Z_G - Z_R}{Z_G + Z_R} \tag{3.24}$$

is the reflection coefficient of the generator port (with respect to the reference impedance Z_R) and c is the impressed wave of the generator.

The reflected wave and the incident wave at the load port are related by the equation

$$b_L = S_L \, a_L \tag{3.25}$$

where

$$S_L = \frac{Z_L - Z_R}{Z_L + Z_R} \tag{3.26}$$

is the reflection coefficient of the load port (with respect to the reference impedance Z_R). If the reference impedances of the connected ports j and k are real, positive, and equal to each other, the ingoing and outgoing waves at these ports must satisfy the relations

$$a_j = b_k \quad \text{and} \quad a_k = b_j \tag{3.27}$$

For convenience, in further investigations the input port of the overall circuit will be referenced as port number 1. For the same reason the output port will be referenced by 2. Of course it does not mean that these ports in any circuit have to be numbered or are numbered as 1 and 2. This convention will be used in the entire book.

Returning to the discussion of input port reflection coefficient, for the generator port and the input port we have

$$a_G = b_1 \quad \text{and} \quad a_1 = b_G \tag{3.28}$$

and for the output port and the load port we have

$$a_L = b_2 \quad \text{and} \quad a_2 = b_L \tag{3.29}$$

According to relations (3.28) and (3.29) the *input port reflection coefficient* is calculated as

$$\Gamma_{IN} = \frac{b_1}{a_1} = \frac{a_G}{a_1} \tag{3.30}$$

The complex impedance looking into the input port is given as

$$Z_{IN} = Z_R \frac{1 + \Gamma_{IN}}{1 - \Gamma_{IN}} = Z_R \frac{1 + \dfrac{a_G}{a_1}}{1 - \dfrac{a_G}{a_1}} = Z_R \frac{a_1 + a_G}{a_1 - a_G} \tag{3.31}$$

3.2.2 Output Port Reflection Coefficient

The output port reflection coefficient Γ_{OUT} and the complex impedance Z_{OUT} seen at the output port of the circuit is calculated in the same way as Γ_{IN} and Z_{IN} but with the assumption that the network is excited at the output port (the impressed wave c is transferred to the load port), as is shown in Figure 3.7.

The output port reflection coefficient Γ_{OUT} is computed as

$$\Gamma_{OUT} = \frac{b_2}{a_2} = \frac{a_L}{a_2} \tag{3.32}$$

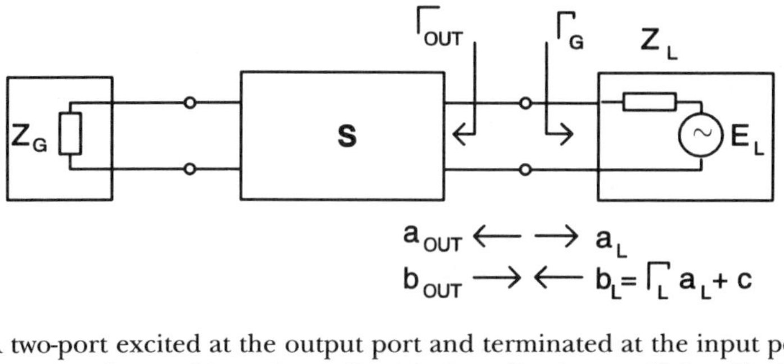

Figure 3.7 A two-port excited at the output port and terminated at the input port by signal generator internal impedance Z_G with the reflection coefficient Γ_G. The network model for computing output port reflection coefficient Γ_{OUT}. Γ_{OUT} is a function of **S** and Γ_G.

The output impedance seen at the output port of the network is then

$$Z_{OUT} = Z_R \frac{1 + \Gamma_{OUT}}{1 - \Gamma_{OUT}} = Z_R \frac{1 + a_L/a_2}{1 - a_L/a_2} = Z_R \frac{a_2 + a_L}{a_2 - a_L} \tag{3.33}$$

3.2.3 Overall Scattering Parameters

The overall scattering parameters of the network related to the input and output ports are calculated according to the network models given in Figures 3.8 and 3.9.

According to Figure 3.6 the matched generator ($\Gamma_G = 0$, $Z_G = Z_R$) with impressed wave $c = 1$ is connected to the input port. The matched load ($\Gamma_L = 0$, $Z_L = Z_R$) is connected to the output port. In such a case the transmission coefficient S_{21} (between the input and output ports) coincides with the wave a_L incident at the load port ($b_2 = a_L$)

$$S_{21} = \frac{b_2}{a_1} = \frac{a_L}{a_1}\bigg|_{a_1=1} = a_L \tag{3.34}$$

The condition $a_1 = 1$ is imposed by the generator connected to the input port with $c = 1$.

The reflection scattering coefficient S_{11} of the overall circuit at its input port coincides with the wave a_G incident at the generator port ($b_1 = a_G$) as

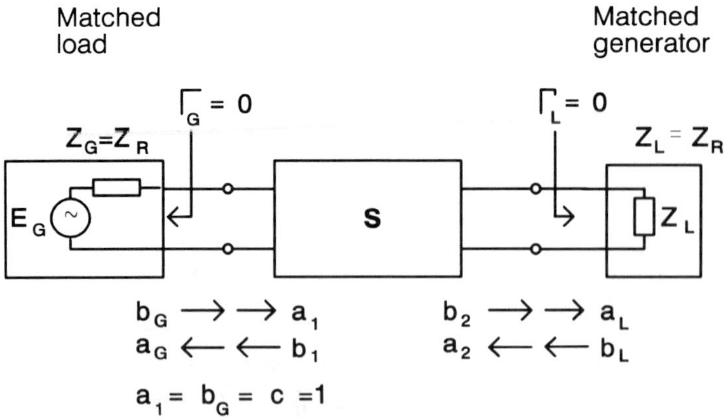

Figure 3.8 A network model for computing S_{11} and S_{21} scattering parameters. The internal impedance Z_G of the signal generator exciting the network at the input port as well as the load impedance Z_L at the output port must be equal to the reference impedance Z_R of network ports.

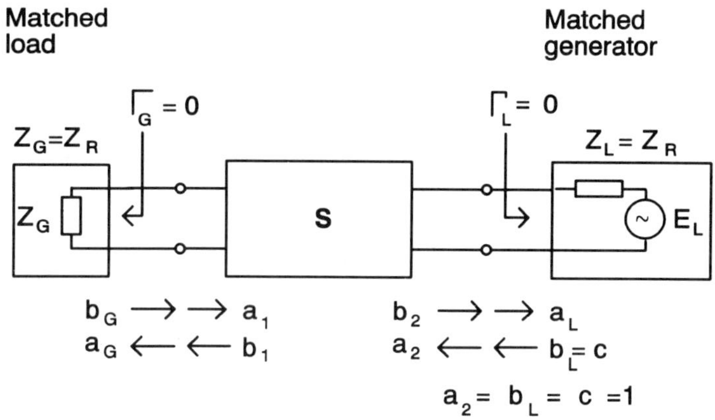

Figure 3.9 A network model for computing S_{22} and S_{12} scattering parameters. The internal impedance Z_L of the signal source exciting the network at the output port as well as the load impedance Z_G at the input port must be equal to the reference impedance Z_R of network ports.

$$S_{11} = \frac{b_1}{a_1} = \frac{a_G}{a_1}\bigg|_{a_1=1} = a_G \tag{3.35}$$

The remaining two scattering matrix parameters are calculated in the same way as S_{21} and S_{11} but with the assumption that the impressed wave $c = 1$ is transferred to the load port, as is shown in Figure 3.9.

This time we have

$$S_{12} = \frac{b_1}{a_2} = \frac{a_G}{a_2}\bigg|_{a_2=1} = a_G \tag{3.36}$$

and

$$S_{22} = \frac{b_2}{a_2} = \frac{a_L}{a_2}\bigg|_{a_2=1} = a_L \tag{3.37}$$

3.2.4 Stability Considerations for Active Networks

Besides gain and bandwidth, stability is one of the most important parameters in active circuit design. Designing a transistor amplifier without considering its stability may lead to useless results. A reliable technique for predicting and assuring stability of active networks uses scattering parameters.

3.2.4.1 Stability Criteria

There are two traditional expressions used when considering active network stability: conditional stability and unconditional stability.

It is said that a two-port is unconditionally stable if the real parts of its input and output impedances remain positive for all passive load and generator impedances.

A circuit is conditionally stable if the real part of its input and output impedances become negative for some positive real (passive) load and generator impedances.

It is important to note that these two conditions (for the input port and for the output port) apply at one specific frequency. As a matter of fact, the conditions that will be discussed will have to be investigated at many frequencies to ensure broadband stability.

The input port and the output port reflection coefficients of a two-port are (Figure 3.10)

$$\Gamma_{IN} = S_{11} + \frac{S_{21}S_{12}\Gamma_L}{1 - S_{22}\Gamma_L} \tag{3.38}$$

$$\Gamma_{OUT} = S_{22} + \frac{S_{21}S_{12}\Gamma_G}{1 - S_{11}\Gamma_G} \tag{3.39}$$

where Γ_G and Γ_L are the reflection coefficients of the generator port and load port, respectively.

The condition of the passiveness of input and output impedances for all possible combinations of the generator and load impedances is identical to

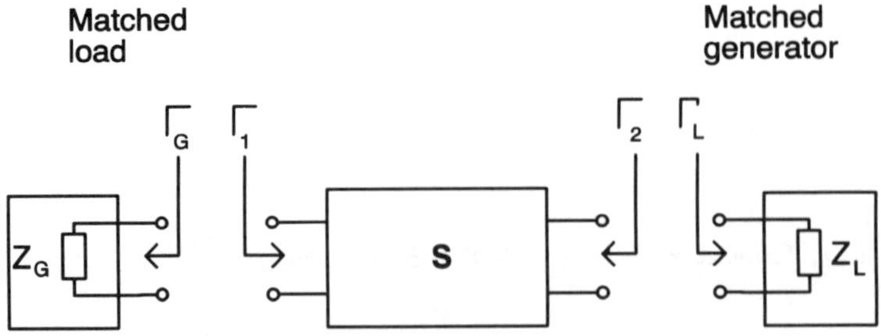

Figure 3.10 A network model for stability considerations of active networks. Input port reflection coefficient Γ_{IN} is a function of **S** and Γ_L. Output port reflection coefficient Γ_{OUT} is a function of **S** and Γ_G.

$$\left| S_{11} + \frac{S_{21}S_{12}\Gamma_L}{1 - S_{22}\Gamma_L} \right| < 1 \quad \text{for all } |\Gamma_L| < 1 \tag{3.40}$$

and

$$\left| S_{22} + \frac{S_{21}S_{12}\Gamma_G}{1 - S_{11}\Gamma_G} \right| < 1 \quad \text{for all } |\Gamma_G| < 1 \tag{3.41}$$

The following necessary and sufficient conditions for unconditional stability of a two-port circuit are a consequence of (3.40) and (3.41) [7]

$$K = \frac{1 + |\Delta|^2 - |S_{11}|^2 - |S_{22}|^2}{2\,|S_{12}|\,|S_{21}|} > 1 \tag{3.42}$$

and

$$|\Delta| = |S_{11}S_{22} - S_{12}S_{21}| < 1 \tag{3.43}$$

Many active devices, such as MESFETs and BJTs, are potentially unstable over a certain frequency range. In such cases it is necessary to know generator and load impedances (reflection coefficients) that correspond to the unstable operation of the device. If the magnitude of Γ_{IN} is set to one, a boundary between the stable and unstable regions on the load reflection coefficient plane may be determined. Solving $|\Gamma_{IN}| = 1$ for all possible values of Γ_L yields the stability circle on the load plane. The center and the radius of the stability circle on the load plane are given by [7, 8]

$$\text{center} \quad C_L = \frac{(S_{22} - \Delta\, S_{11}^*)^*}{|S_{22}|^2 - |\Delta|^2} \tag{3.44}$$

$$\text{radius} \quad r_L = \left| \frac{S_{12}S_{21}}{|S_{22}|^2 - |\Delta|^2} \right| \tag{3.45}$$

where $\Delta = S_{11}S_{22} - S_{12}S_{21}$.

Similarly, solving $|\Gamma_{OUT}| = 1$ for all possible values of Γ_G yields a stability circle on the signal generator reflection coefficient plane. The center and the radius of the stability circle on the generator plane are given by [7, 8]

$$\text{center} \quad C_G = \frac{(S_{11} - \Delta\, S_{22}^*)^*}{|s_{11}|^2 - |\Delta|^2} \tag{3.46}$$

$$\text{radius} \quad r_L = \left| \frac{S_{12}S_{21}}{|S_{11}|^2 - |\Delta|^2} \right| \tag{3.47}$$

where $\Delta = S_{11}S_{22} - S_{12}S_{21}$ and the asterisk represents the complex conjugate operation.

The stability circles can be plotted directly on the Smith chart; they separate the load or signal generator reflection coefficient plane into stable and potentially unstable regions. To determine which area of the chart represents the stable region, it is necessary to perform certain considerations. In the investigations to follow, we will consider the load plane and the load stability circle. Let $Z_L = Z_R$, where Z_R is the real reference impedance of ports of the two-port. Then $\Gamma_L = 0$, and on the Γ_L plane this value of reflection coefficient corresponds to the center of the Smith chart. According to (3.38), $\Gamma_L = 0$ corresponds to $\Gamma_{IN} = S_{11}$ and if $|S_{11}| < 1$, then also $|\Gamma_{IN}| < 1$. In consequence this means that if the center of the Smith chart lies outside the stability circle, then the whole area of the chart outside the load stability circle represents the stable region for Γ_L. This case is illustrated in Figure 3.11(a).

If $|S_{11}| > 1$ then for $\Gamma_L = 0$ also $|\Gamma_{IN}| > 1$. This means that in the case when the center of Smith chart lies outside the stability circle then only the area inside the stability circle represents the stable region for Γ_L. Figure 3.11(b) illustrates this case of stability region.

Similar considerations could be performed for the signal generator port.

Stability considerations of a network are very important parts of microwave amplifier or generator design procedure. The stability circles plotted in the Γ_L plane and Γ_G plane for these frequencies where $K < 1$ facilitate the design of the matching circuit or feedback circuit.

It should be mentioned here also that the stability conditions presented in this

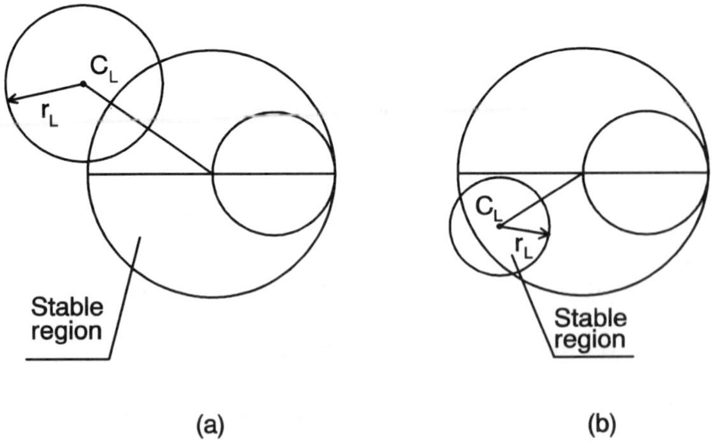

Figure 3.11 Stability of a two-port network considered in the Γ_L plane of the Smith chart. (a) Stability circle and stable region for a network with $|S_{11}| < 1$. (b) Stability circle and stable region for a circuit with $|S_{11}| > 1$.

section are considered to be sufficient for a single transistor chip treated as a two-port circuit. For the more general case of cascaded amplifiers or a single transistor embedded in a passive network containing reactive elements, we must be very careful in stability considerations.

3.2.5 Power Gain

With reference to the general two-port circuit shown in Figure 3.12, the power gain is defined as

$$\text{Power Gain} \quad G = \frac{P_L}{P_{IN}} \tag{3.48}$$

where P_L is the power delivered to the load and P_{IN} is the power supplied to the input port of the circuit. In terms of the incident and reflected waves at circuit ports (Figure 3.12), the power gain is

$$G = \frac{|a_L|^2 - |b_L|^2}{|a_1|^2 - |b_1|^2} = \frac{|a_L|^2 - |a_2|^2}{|a_1|^2 - |a_G|^2}$$

$$= \frac{|a_L|^2 (1 - |S_L|^2)}{|a_1|^2 - |a_G|^2} \tag{3.49}$$

3.2.6 Transducer Power Gain

Transducer power gain is defined as

$$\text{Transducer Power Gain} \quad G_T = \frac{P_L}{P_{GA}} \tag{3.50}$$

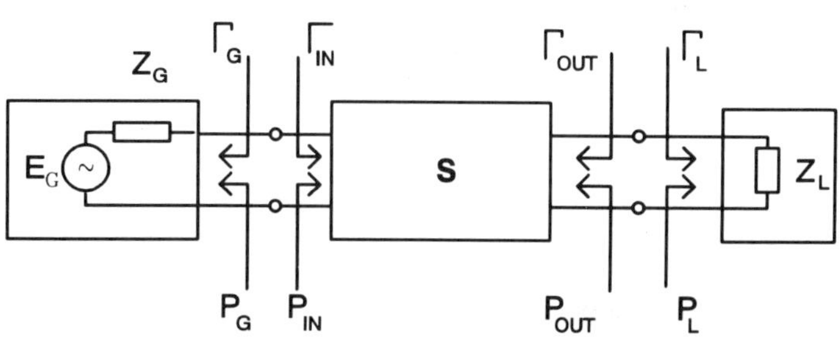

Figure 3.12 A network model for power gain computation. Power gain is a function of **S** and Γ_L.

where P_{GA} is the power available from the signal generator under matching conditions. Clearly, the transducer power gain is a function of circuit-scattering parameters, the load port reflection coefficient (load impedance), and the signal generator port reflection coefficient (internal impedance of the signal generator).

To express G_T in terms of incident and reflected waves at network ports we must first determine the available power of the signal generator in terms of wave variables. Available power is the maximum power that may be delivered to the load by the signal generator. From basic circuit theory it is known that $P_L = P_{GA}$ when the load impedance Z_L equals the complex conjugate of the generator internal impedance; that is, when

$$Z_L = Z_G^* \tag{3.51}$$

In the reflection coefficient domain, the equivalent to (3.51) condition has the form

$$S_L = S_G^* \tag{3.52}$$

Using the circuit model shown in Figure 3.13, it is easy to evaluate the available power of the signal generator. It is equal to

$$P_{GA} = \frac{|c|^2}{1 - |S_G|^2} \tag{3.53}$$

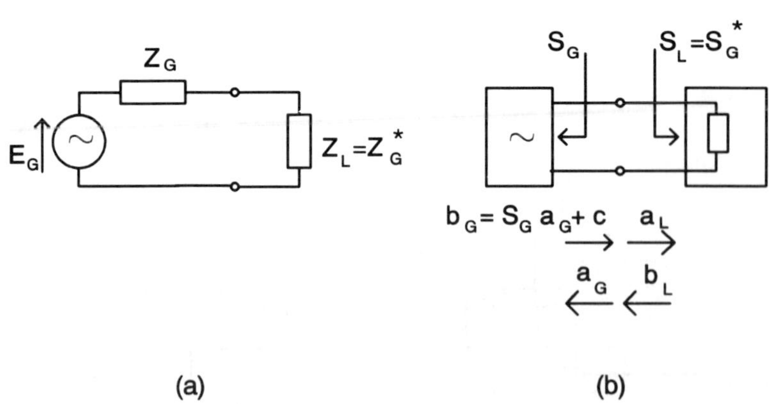

Figure 3.13 Conjugate match of the signal generator. (a) Load impedance Z_L matched to the voltage source with internal impedance Z_G, $Z_L = Z_G^*$. (b) Load one-port with reflection coefficient S_L matched to signal generator with internal reflection coefficient S_G, $S_L = S_G^*$.

where c is the impressed wave and S_G is the reflection coefficient of the signal generator.

Because real power P_L delivered to the load is

$$P_L = |a_L|^2 - |b_L|^2 = |a_L| (1 - |S_L|^2) \tag{3.54}$$

using (3.50), (3.53), and (3.55) we have

$$G_T = \frac{|a_L|^2 (1 - |S_L|^2)(1 - |S_G|^2)}{|c|^2} \tag{3.55}$$

For numerical convenience, the impressed wave c of the signal generator is always set to 1. Then G_T is computed as

$$G_T = |a_L|^2 (1 - |S_L|^2)(1 - |S_G|^2) \tag{3.56}$$

3.2.7 Available Power Gain

Available power gain is defined as

$$\text{Available Power Gain} \quad G_A = \frac{P_{OUTA}}{P_{GA}} \tag{3.57}$$

where P_{OUTA} is the power available at the output port of the circuit and P_{GA} is the power available from the signal generator. P_{OUTA} is the power delivered to the output port load under matching conditions, that is, when $\Gamma_L = \Gamma^*_{OUT}$. The available power gain is a function of the circuit-scattering parameters and of the signal generator port reflection coefficient (internal impedance of the signal generator). It is, however, independent of the load impedance.

Referring to Figures 3.12 and 3.7 it may be proved that the available power gain equals [7, 8]

$$G_A = \frac{|S_{21}|^2 (1 - |\Gamma_G|^2)}{1 - |S_{22}|^2 + |\Gamma_G|^2 (|S_{11}|^2 - |\Delta|^2) - 2\,\text{Re}\{C_1 \Gamma_G\}} \tag{3.58}$$

where $\Delta = S_{11} S_{22} - S_{12} S_{21}$ and $C_1 = S_{11} - \Delta S^*_{22}$.

3.2.8 Simultaneous Conjugate Match and Maximum Power Gain

If the stability factor $K > 1$ and $|\Delta| < 1$, the device is unconditionally stable and the maximum possible gain can be achieved by conjugate matching the input and the output of the device.

The conditions for simultaneous match are

$$\Gamma_{IN} = S_{11} + \frac{S_{12}\, S_{21}\, \Gamma_L}{1 - S_{22}\, \Gamma_L} = \Gamma_G^* \tag{3.59}$$

and

$$\Gamma_{OUT} = S_{22} + \frac{S_{12}\, S_{21}\, \Gamma_G}{1 - S_{22}\, \Gamma_G} = \Gamma_L^* \tag{3.60}$$

It may be proven that the only useful solution of (3.59) and (3.60) for Γ_G and Γ_L is [8]

$$\Gamma_{GM} = \frac{C_1^*}{|C_1|}\left[\frac{B_1}{2|C_1|} - \sqrt{\frac{B_1^2}{|2C_1|^2} - 1}\right] \tag{3.61}$$

and

$$\Gamma_{LM} = \frac{C_2^*}{|C_2|}\left[\frac{B_2}{2|C_2|} - \sqrt{\frac{B_2^2}{|2C_2|^2} - 1}\right] \tag{3.62}$$

where

$$C_1 = S_{11} - \Delta S_{22}^* \tag{3.63}$$

$$B_1 = 1 - |S_{22}|^2 + |S_{11}|^2 - |\Delta|^2 \tag{3.64}$$

$$C_2 = S_{22} - \Delta\, S_{11}^* \tag{3.65}$$

$$B_2 = 1 - |S_{11}|^2 + |S_{22}|^2 - |\Delta|^2 \tag{3.66}$$

Under simultaneous conjugate match at the input and output ports of the two-port, power gain, available power gain, and transducer power gain are all the same ($G = G_A = G_T$) and equal to maximum possible gain G_{MAX}, which is given as [7, 8]

$$G_{MAX} = \left|\frac{S_{21}}{S_{12}}\right|(K \pm \sqrt{K^2 - 1}) \tag{3.67}$$

where $K > 1$ and the minus sign is used in front of the radical if the quantity B_1 (3.64) is positive. If B_1 is negative the plus sign should be used in (3.67) to compute G_{MAX}.

3.2.9 Constant Power Gain Circles

When the two-port circuit is potentially unstable and $K < 1$, G_{MAX} does not exist. Designing amplifiers with potentially unstable transistors requires the use of power gain or available power gain. It can be proved that power gain expressed by scattering parameters is equal to [7, 8]

$$G = \frac{|S_{21}|^2 (1 - |\Gamma_L|^2)}{1 - |S_{11}|^2 + |\Gamma_L|^2 (|S_{21}|^2 - |\Delta|^2) - 2\,\mathrm{Re}\{C_2\Gamma_L\}} \tag{3.68}$$

where $C_2 = S_{22} - \Delta S_{11}^*$.

It can be shown that the equation expressing normalized power gain $g = G/|S_{21}|^2$ represents, for a given g, a constant gain circle on the load reflection coefficient plane.

This circle is the locus of all Γ_L for which the power gain of the circuit is constant and equal $G = g\,|S_{21}|^2$.

The center C_p and the radius r_p of the constant power gain circle are defined by the equations [7, 8]

$$\text{center} \quad C_p = \frac{g\,(S_{22}^* - \Delta^* S_{11})}{1 + g\,(|S_{22}|^2 - |\Delta|^2)} \tag{3.69}$$

$$\text{radius} \quad r_p = \frac{(1 - 2K\,|S_{12}S_{21}|\,g + |S_{12}S_{21}|^2\,g^2)^{1/2}}{|1 + g\,(|S_{22}|^2 - |\Delta|^2)|} \tag{3.70}$$

3.2.10 Constant Available Power Gain Circles

It can be proved that the available power gain of a two-port network expressed by the scattering parameters of the network and Γ_G is equal to [7, 8]

$$G_A = \frac{|S_{21}|^2 (1 - |\Gamma_G|^2)}{1 - |S_{22}|^2 + |\Gamma_G|^2 (|S_{11}|^2 - |\Delta|^2) - 2\,\mathrm{Re}\{C_1\Gamma_G\}} \tag{3.71}$$

where $C_1 = S_{11} - \Delta S_{22}^*$.

As for the power gain, it can be proved that the normalized available power gain $g_A = G_A/|S_{21}|^2$ represents, for a given g_A, a constant available power gain circle on the generator port reflection coefficient plane. The center C_A and the radius r_A of such a circle express the following relations

$$\text{center} \quad C_A = \frac{g_A(S_{11}^* - \Delta^* S_{22})}{1 + g_A(|S_{11}|^2 - |\Delta|^2)} \tag{3.72}$$

$$\text{radius} \quad r_A = \frac{(1 - 2K|S_{12}S_{21}|\, g_A + |S_{12}S_{21}|^2\, g_A^2)^{1/2}}{|1 + g_A(|S_{11}|^2 - |\Delta|^2)|} \tag{3.73}$$

3.2.11 Insertion Loss

The insertion loss of the network is defined as

$$\text{Insertion Loss} \quad \text{IL} = \frac{P_{LO}}{P_L} \tag{3.74}$$

where P_{LO} is the power delivered to the load connected directly to the generator port and P_L is the power delivered to the load with the circuit inserted between the generator and the load. Figure 3.14 illustrates the definition of IL. Because the power delivered to the load connected directly to the generator port is

$$P_{LO} = \frac{|c|^2 (1 - |\Gamma_L|^2)}{|1 - \Gamma_G \Gamma_L|^2} \tag{3.75}$$

and the power delivered to the load of the circuit is

$$P_L = |a_L|^2(1 - |\Gamma_L|^2) = \frac{|c|^2 |S_{21}|^2 (1 - |\Gamma_L|^2)}{|1 - S_{11}\Gamma_G - S_{22}\Gamma_L + \Delta_s \Gamma_G \Gamma_L|^2} \tag{3.76}$$

the insertion loss expressed in terms of two-port scattering parameters and generator and load port reflection coefficients is $(c = 1)$

$$\text{IL} = \frac{|1 - S_{11}\Gamma_G - S_{22}\Gamma_L + \Delta_s \Gamma_G \Gamma_L|^2}{|S_{21}|^2 |1 - \Gamma_G \Gamma_L|^2} \tag{3.77}$$

In (3.76) and (3.77)

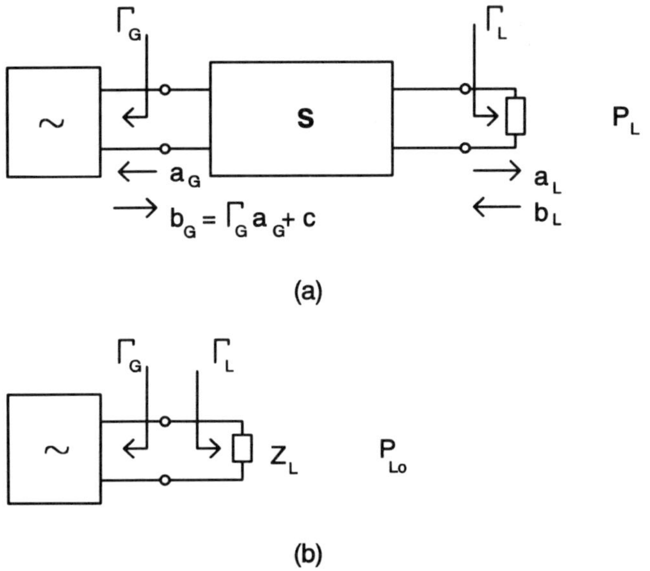

Figure 3.14 A network model for insertion loss computation. (a) A two-port with the scattering matrix **S** inserted between the generator and the load. (b) The load with reflection coefficient Γ_L directly connected to the generator port.

$$\Delta_s = S_{11}\, S_{22} - S_{21}\, S_{21} \tag{3.78}$$

3.2.12 Group Delay

For microwave circuits described by scattering parameters, the group delay of a circuit defines the relation

$$\tau_D = -\frac{\partial\, \phi_{21}}{\partial\, \omega} \tag{3.79}$$

where $\Phi_{21} = \arg\{S_{21}\}$ and S_{21} is the overall transmission scattering parameter between the input port 1 and the output port 2. The group delay may be computed by means of the incremental ratio approximation

$$\tau_D = -\frac{\Delta\, \phi_{21}}{\Delta\, \omega} \tag{3.80}$$

Obviously, the smaller is the frequency step $\Delta\omega$, the better is the approximation of the group delay τ_D by (3.80).

A more accurate and more precise numerical way of group delay computation is based on the adjoint network method. It will be discussed in Section 5.4 of this book.

3.2.13 Voltage Transfer Gain

Voltage transfer gain is defined as (see Figure 3.15)

$$\text{Voltage Transfer Gain} \quad G_v = \frac{V_L}{E_G} \tag{3.81}$$

where V_L is the voltage at the load and E_G is the signal generator voltage.

Because $E_G = V_1 + Z_G I_1$, using (3.3) and (3.4) we have

$$E_G = \sqrt{Z_R}\,(a_1 + b_1) + \frac{Z_G}{\sqrt{Z_R}}(a_1 - b_1) \tag{3.82}$$

and

$$V_L = \sqrt{Z_R}\,(a_L + b_L) \tag{3.83}$$

From the above equations, the voltage transfer gain expressed in terms of wave variables is equal to

$$G_v = \frac{\sqrt{Z_R}\,(a_L + b_L)}{\sqrt{Z_R}\,(a_1 + b_1) + \frac{Z_G}{\sqrt{Z_R}}(a_1 - b_1)}$$

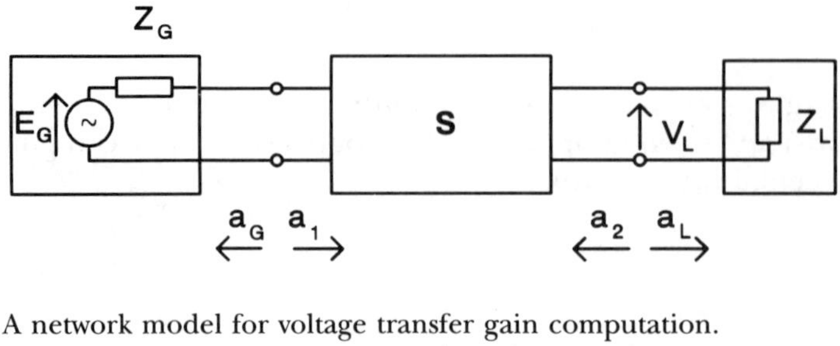

Figure 3.15 A network model for voltage transfer gain computation.

$$= \frac{Z_R (a_L + a_2)}{Z_R (a_1 + a_G) + Z_G (a_1 - a_G)} \tag{3.84}$$

where Z_R is the real and positive reference impedance common to all circuit ports and Z_G is the internal impedance of the signal generator. The generator impedance Z_G may be computed as

$$Z_G = Z_R \frac{1 + \Gamma_G}{1 - \Gamma_G} = Z_R \frac{1 + S_G}{1 - S_G} \tag{3.85}$$

Amending (3.84) into the form

$$G_v = \frac{Z_R (1 + \Gamma_L)}{Z_R (1 + \Gamma_{IN}) + Z_G (1 - \Gamma_{IN})} \cdot \frac{b_2}{a_1} \tag{3.86}$$

and using the relation

$$\frac{b_2}{a_1} = \frac{S_{21}}{1 - S_{22}\Gamma_L} \tag{3.87}$$

we get the final relation used to compute the transfer voltage gain

$$G_v = \frac{Z_R (1 + \Gamma_L)}{Z_R (1 + \Gamma_{IN}) + Z_G (1 - \Gamma_{IN})} \cdot \frac{S_{21}}{1 - S_{22}\Gamma_L} \tag{3.88}$$

Transmission phase shift Φ is the argument of G_v

$$\Phi = \arg \left\{ \frac{E_g}{U} \right\} = -\arg \{G_v\} \tag{3.89}$$

and is calculated using (3.84).

References

[1] Monaco, V. A., and P. Tiberio, "On the Transformation of a Lumped Element Linear Network Into a Circuit Composed of Multiports," *Alta Frequenca,* Vol. 39, November 1970, pp. 1013–1014.

[2] Petenpaul, E., H. Kapusta, A. Weisegerber, H. Mampe, J. Lugins, and I. Wolf, "CAD Models of Lumped Elements on GaAs Up To 18 GHz," *IEEE Trans. Microwave Theory Tech.,* Vol. 36, No. 2, February 1988, pp. 294–304.

[3] Dobrowolski, J. A., *Introduction to Computer Methods for Microwave Circuit Analysis and Design*, Norwood, MA: Artech House, 1991.
[4] Monaco, V. A., and P. Tiberio, "Computer-Aided Analysis of Microwave Circuits," *IEEE Trans. Microwave Theory Tech.*, Vol. MTT-22, March 1974, pp. 249–263.
[5] Gupta, K. C., R. Garg, and R. Chadha, *Computer-Aided Design of Microwave Circuits*, Dedham, MA: Artech House, 1981.
[6] Monaco, V. A., and P. Tiberio, "Automatic Scattering Matrix Computation of Microwave Circuits," *Alta Frequenca*, Vol. 39, February 1970, pp. 59–64.
[7] Bodway, G. E., "Two Port Power Flow Analysis Using Generalized Scattering Parameters," *Microwave Journal*, Vol. 10, May 1967, pp. 61–69.
[8] S-Parameter Design, Hewlett-Packard, Application Note 154, April 1972.

CHAPTER 4

Noise Analysis of Microwave Networks Using Noise Waves

There are many ways to characterize noise in linear microwave networks. The most commonly used representations of noise generated in networks are equivalent voltage or current sources or a combination of both. Impedance and admittance matrix representations, chain matrix representation, and a few others are used in CAD analysis of noisy networks [1].

At microwave frequencies, however, a wave representation of noise is advantageous. It allows the use of scattering matrices for noise computations, leading to advantages in CAD of microwave networks. The power of the wave representation of noise lies in its compatibility with the scattering matrix description of microwave networks. This permits noise analysis problems to be formalized and solved using scattering parameters. Noise analysis presented in this chapter is based on the connection scattering matrix formalism used for circuit description. The applied noise analysis method is fully compatible with conventional circuit analysis for sinusoidal excitation. The noise analysis method facilitates the computation of noise power sensitivities that may be used for noise figure minimization using the optimization method.

4.1 NOISE WAVE REPRESENTATION OF NOISY MICROWAVE NETWORKS

In the noise wave representation a circuit element's noise is described using waves that emanate from its ports [2].

As it is shown in Figure 4.1, noise generated in a two-port element is represented by two mutually correlated noise wave sources c_1 and c_2, one source at each port. The linear matrix equation describing the noisy two-port is

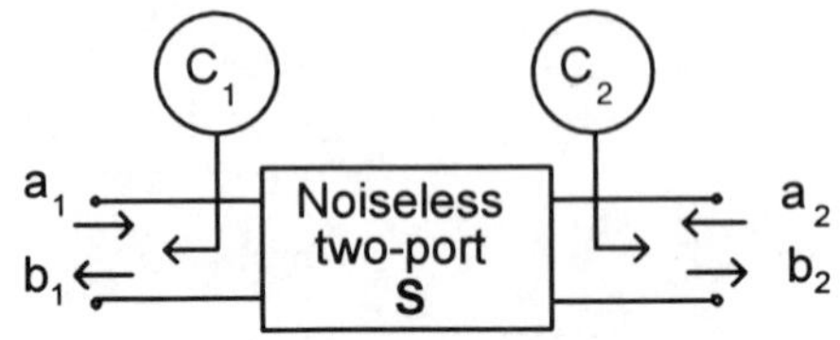

Figure 4.1 Noise wave and scattering matrix representation of a noisy two-port. The noise wave sources c_1 and c_2 represent noise generated in a noisy two-port.

$$\begin{bmatrix} b_1 \\ b_2 \end{bmatrix} = \begin{bmatrix} S_{11} & S_{12} \\ S_{21} & S_{22} \end{bmatrix} \begin{bmatrix} a_1 \\ a_2 \end{bmatrix} + \begin{bmatrix} c_1 \\ c_2 \end{bmatrix} \tag{4.1}$$

The wave representation of noise in a linear two-port is originally due to Bauer and Rothe [3]. The ingoing and outgoing noise waves a_i and b_i at the ith port are defined in terms of port terminal noise voltage V_i and noise current I_i and a complex reference (normalization) impedance Z_{Ri}

$$a_i = \frac{V_i + Z_{Ri} I_i}{2\sqrt{\mathrm{Re}\{Z_{Ri}\}}} \tag{4.2}$$

and

$$b_i = \frac{V_i - Z_{Ri}^* I_i}{2\sqrt{\mathrm{Re}\{Z_{Ri}\}}} \tag{4.3}$$

where the asterisk indicates the complex conjugate and Re the real part. In (4.2) and (4.3), Z_R is a port reference impedance that may be the same as that used to define the power waves and scattering parameters of the circuit. Equations (4.2) and (4.3) are identical to the standard definition of power waves for deterministic signals [4].

Variables a_i and b_i are the complex amplitudes of the ingoing and outgoing noise waves, respectively, while $|a_i|^2$ and $|b_i|^2$ are the corresponding noise powers carried by these waves.

It is convenient to assume that the reference impedance Z_R in (4.2) and (4.3) and the reference impedance Z_R used in (2.1) and (2.2) are always real and equal to each other.

The noise wave sources c_1 and c_2 representing noise generated in a two-port are time-varying complex variables characterized by a correlation matrix $\mathbf{C}_S$ given by

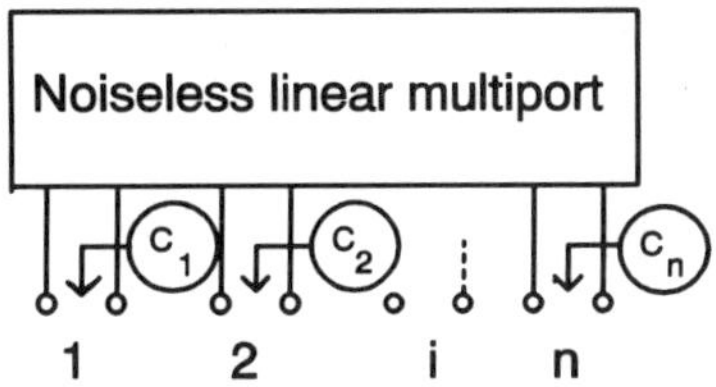

Figure 4.2 Noise wave representation of a noisy multiport. The nose wave sources c_i, $i = 1, 2, \ldots, n$, are characterized by a correlation matrix $\mathbf{C}_s$.

$$\mathbf{C}_S = \overline{\begin{bmatrix} c_1 \\ c_2 \end{bmatrix}\begin{bmatrix} c_1 \\ c_2 \end{bmatrix}^+} = \begin{bmatrix} \overline{c_1 c_1^*} & \overline{c_1 c_2^*} \\ \overline{c_2 c_1^*} & \overline{c_2 c_2^*} \end{bmatrix} = \begin{bmatrix} \overline{|c_1|^2} & \overline{c_1 c_2^*} \\ \overline{c_2 c_1^*} & \overline{|c_2|^2} \end{bmatrix} \tag{4.4}$$

where the overbars indicate time averaging, + Hermitian conjugate, and * complex conjugate. The diagonal terms of $\mathbf{C}_S$ give the noise power deliverable to the matched terminations of ports in a 1-Hz bandwidth. The off-diagonal terms are correlation products. The noise wave correlation matrices $\mathbf{C}_S$ are Hermitian matrices. Their components are referred to as noise wave parameters.

In the general multiport case, shown in Figure 4.2, noise generated in a network is represented by mutually correlated noise wave sources, one source at each port. The matrix equation for the noise wave representation of a linear multiport is

$$\mathbf{b} = \mathbf{S}\,\mathbf{a} + \mathbf{c} \tag{4.5}$$

where $\mathbf{S}$ denotes the scattering matrix of the multiport, $\mathbf{a} = [a_1\ a_2 \cdots a_n]^T$ is a vector of the port incoming noise waves, $\mathbf{b} = [b_1\ b_2 \cdots b_n]^T$ is a vector of the port outgoing noise waves, and $\mathbf{c} = [c_1\ c_2 \cdots c_n]^T$ is a vector of the port equivalent noise wave sources.

The multiport noise wave correlation matrix $\mathbf{C}_S$ of a network element is given by

$$\mathbf{C}_S = \overline{\mathbf{c}\,\mathbf{c}^+} \tag{4.6}$$

Scattering parameters as well as noise parameters of a linear multiport are frequency dependent. The full characterization of signal and noise properties of a multiport requires knowledge of the scattering matrix and of the noise wave correlation matrix at each frequency of interest.

4.2 OTHER NOISE REPRESENTATIONS OF NOISY NETWORKS AND THEIR TRANSFORMATIONS TO NOISE WAVE PARAMETERS

A physically significant description of noise sources existing in the equivalent circuits of noisy two-ports is given by their self- and cross-power spectral densities, which are defined as the Fourier transform of their auto- and cross-correlation functions. Arranging these spectral densities in matrix form leads to the so-called correlation matrices [5, 6].

Noise correlation matrices are Hermitian matrices because

$$\text{Im } C_{11} = \text{Im } C_{22} = 0 \quad \text{and} \quad C_{12} = C_{21}^{*} \tag{4.7}$$

Hence, noise properties of linear noisy two-ports are fully described by four real numbers

$$C_{11},\ C_{22},\ \text{Re } C_{12},\ \text{and Im } C_{12} \tag{4.8}$$

Another very important property of these matrices is that they are positive semidefinite; that is,

$$C_{11} \geq 0 \qquad (C_{22} \geq 0) \tag{4.9}$$

and

$$\det \mathbf{C}_S = C_{11}\ C_{22} - |\ C_{12}\ |^2 \geq 0 \tag{4.10}$$

Noise correlation matrices have been known for years [7, 8], but in 1975 K. Reiss [6] and H. Hillbrandt and P. Russer [7] proved their importance in the analysis of linear noisy networks.

4.2.1 Chain Matrix Noise Representation

Figure 4.1 illustrates the Rothe and Dahlke equivalent circuit of a noisy two-port [2]. It uses two external, input port noise sources E_N and J_N. Because the primary physical noise sources existing inside the circuit contribute to the equivalent external noise sources, E_N and J_N are not statistically independent. The equations describing this noise representation have the form

$$\begin{bmatrix} V_1 \\ I_1 \end{bmatrix} = \begin{bmatrix} A & B \\ C & D \end{bmatrix} \begin{bmatrix} V_2 \\ -I_2 \end{bmatrix} + \begin{bmatrix} E_N \\ J_N \end{bmatrix} \tag{4.11}$$

where A_1 B, C, D are the chain matrix parameters.

The noise correlation matrix for chain matrix representation is

$$\mathbf{C}_A = \overline{\begin{bmatrix} E_N \\ J_N \end{bmatrix} \begin{bmatrix} E_N \\ J_N \end{bmatrix}^+} = \begin{bmatrix} \overline{E_N E_N^*} & \overline{E_N J_N^*} \\ \overline{J_N E_N^*} & \overline{J_N J_N^*} \end{bmatrix} \tag{4.12}$$

where + denotes the conjugate transpose and * denotes the complex conjugate.

Noise parameters for the chain matrix representation of a noisy two-port are R_n, g_n, and ρ, where

$$R_n = \frac{\overline{|E_N|^2}}{4kT_0 df} \qquad g_n = \frac{\overline{|J_N|^2}}{4kT_0 df} \tag{4.13}$$

$$\rho = \frac{\overline{E_N^* J_N}}{\sqrt{\overline{|E_N|^2}\,\overline{|J_N|^2}}} \tag{4.14}$$

The noise correlation matrix for the scattering matrix representation has the form given by (4.4).

The interrelation of the chain matrix and the scattering matrix representations is very important. It comes from the following considerations and description given in Figure 4.3. The circuit in Figure 4.3(a) fulfills the relations

$$V_1 = V_1' + E_N \quad \text{and} \quad I_1 = I_1' + J_N \tag{4.15}$$

For the equivalent scattering matrix description given in Figure 4.4(b), the appropriate relations are

$$\begin{bmatrix} S_{11} & S_{12} \\ S_{21} & S_{22} \end{bmatrix} \begin{bmatrix} a_1' \\ a_2 \end{bmatrix} = \begin{bmatrix} b_1' \\ b_2 \end{bmatrix} + \begin{bmatrix} c_1 \\ c_2 \end{bmatrix} \tag{4.16}$$

where

$$a_1' = \frac{V_1' + Z_R I_1'}{2\sqrt{Z_R}} \qquad a_2 = \frac{V_2 + Z_R I_2}{2\sqrt{Z_R}} \tag{4.17}$$

$$b_1' = \frac{V_1' - Z_R I_1'}{2\sqrt{Z_R}} \qquad b_2 = \frac{V_2 - Z_R I_2}{2\sqrt{Z_R}} \tag{4.18}$$

Using relations (4.15), (4.17), and (4.18) we obtain

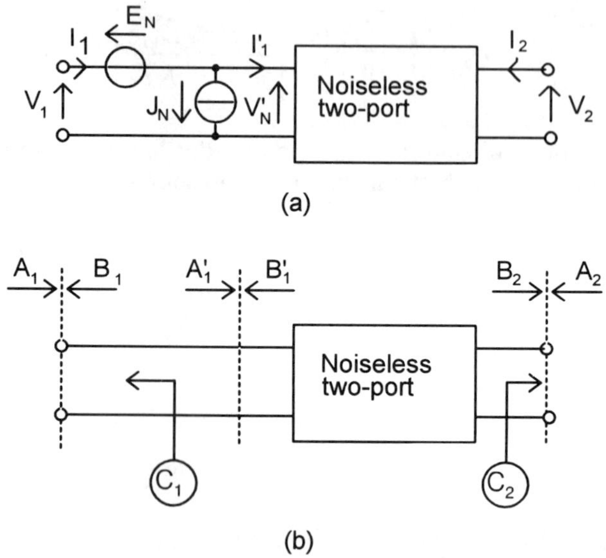

Figure 4.3 The schematic representation of noise properties of a noisy two-port element: (a) using chain matrix and equivalent noise voltage and current sources, both located at the input port; (b) using scattering matrix and noise wave sources located at the input port and at the output port of the element.

$$a_1' = a_1 - \frac{E_N + Z_R J_N}{2\sqrt{Z_R}} \quad \text{and} \quad b_1' = b_1 - \frac{E_N - Z_R J_N}{2\sqrt{Z_R}} \tag{4.19}$$

Then, after substituting the above dependencies into (4.16), we derive formulas for the noise wave sources c_1 and c_2 as functions of E_N and J_N:

$$c_1 = \frac{E_N - Z_R J_N}{2\sqrt{Z_R}} - S_{11}\frac{E_N + Z_R J_N}{2\sqrt{Z_R}} \tag{4.20}$$

$$c_2 = -S_{21}\frac{E_N + Z_R J_N}{2\sqrt{Z_R}} \tag{4.21}$$

To derive relations between the noise correlation matrices for the chain matrix and for the scattering matrix representations we write (4.18) in matrix form as

$$\begin{aligned}\begin{bmatrix} c_1 \\ c_2 \end{bmatrix} &= \frac{1}{2\sqrt{Z_R}}\begin{bmatrix} -S_{11} & 1 \\ -S_{21} & 0 \end{bmatrix}\begin{bmatrix} 1 & Z_R \\ 1 & -Z_R \end{bmatrix}\begin{bmatrix} E_N \\ J_N \end{bmatrix} \\ &= \frac{1}{2\sqrt{Z_R}}\mathbf{P}_1\,\mathbf{P}_2\begin{bmatrix} E_N \\ J_N \end{bmatrix}\end{aligned} \tag{4.22}$$

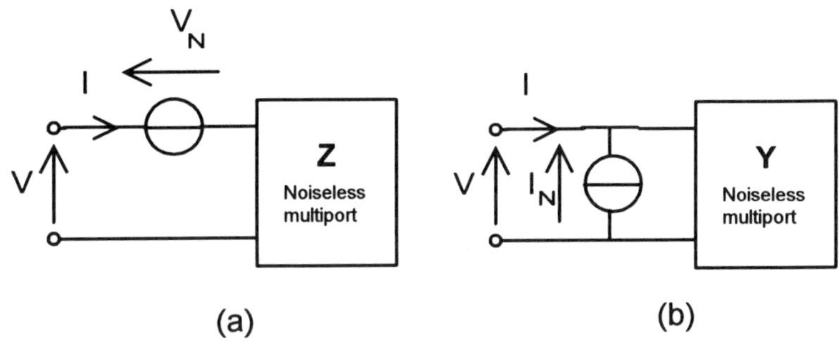

Figure 4.4 (a) Impedance matrix noise representation of a noisy multiport with equivalent noise voltage source at each port. (b) Admittance matrix noise representation of a noisy multiport with equivalent noise current source at each port.

from which we receive relations between both representations

$$\mathbf{C}_S = \overline{\begin{bmatrix} c_1 \\ c_2 \end{bmatrix} \begin{bmatrix} c_1 \\ c_2 \end{bmatrix}^+} = \begin{bmatrix} \overline{c_1 c_1^*} & \overline{c_1 c_2^*} \\ \overline{c_2 c_1^*} & \overline{c_2 c_2^*} \end{bmatrix} = \begin{bmatrix} c_{S11} & c_{S12} \\ c_{S21} & c_{S22} \end{bmatrix}$$

$$= \frac{1}{2\sqrt{Z_R}} \mathbf{P}_1 \, \mathbf{P}_2 \overline{\begin{bmatrix} E_N \\ J_N \end{bmatrix} \begin{bmatrix} E_N \\ J_N \end{bmatrix}^+} \mathbf{P}_2^+ \, \mathbf{P}_1^+ \frac{1}{2\sqrt{Z_R}}$$

$$= \frac{1}{4\, Z_R} \mathbf{P}_1 \, \mathbf{P}_2 \, \mathbf{C}_A \, \mathbf{P}_2^+ \, \mathbf{P}_1^+ \tag{4.23}$$

where $\mathbf{C}_S$ is the noise correlation matrix for the scattering matrix representation and

$$\mathbf{C}_A = \overline{\begin{bmatrix} E_N \\ J_N \end{bmatrix} \begin{bmatrix} E_N \\ J_N \end{bmatrix}^+} = \begin{bmatrix} \overline{E_N E_N^*} & \overline{E_N J_N^*} \\ \overline{E_N J_N^*} & \overline{J_N J_N^*} \end{bmatrix} \tag{4.24}$$

is the noise correlation matrix for the chain matrix representation. According to (4.22) the transformation matrices $\mathbf{P}_1$ and $\mathbf{P}_2$ have the forms

$$\mathbf{P}_1 = \begin{bmatrix} -S_{11} & 1 \\ -S_{21} & 0 \end{bmatrix} \quad \text{and} \quad \mathbf{P}_2 = \begin{bmatrix} 1 & Z_R \\ 1 & -Z_R \end{bmatrix} \tag{4.25}$$

The reverse relation of (4.23) has the form

$$\mathbf{C}_A = 4\ Z_R\ \mathbf{P}_2^{-1}\ \mathbf{P}_1^{-1}\ \mathbf{C}_S\ (P_1^+)^{-1}\ (P_2^+)^{-1} \tag{4.26}$$

For the normalized noise correlation matrices $\mathbf{C}_A$ and $\mathbf{C}_S$ we have

$$\mathbf{C}_S = \frac{1}{Z_R}\ \mathbf{P}_1\ \mathbf{P}_2\ \mathbf{C}_A\ \mathbf{P}_2^+\ \mathbf{P}_1^+ \tag{4.27}$$

and

$$\mathbf{C}_A = Z_R\ \mathbf{P}_2^+\ \mathbf{P}_1^+\ \mathbf{C}_S\ (\mathbf{P}_1^+)^{-1}\ (\mathbf{P}_2^+)^{-1} \tag{4.28}$$

4.2.2 Impedance Matrix and Admittance Matrix Noise Representations

Figure 4.4 presents impedance matrix and admittance matrix noise representations of noisy multiports. The use of multiple noise representations for CAD noise analysis has been shown in [1, 9]. Transformations are used to simplify calculations or to avoid singularities.

Next we give the families of relations describing transformations between wave, impedance, and admittance noise parameters.

Impedance Matrix Representation

Circuit equation

$$\mathbf{v} = \mathbf{Z}\ \mathbf{i} + \mathbf{v}_N \tag{4.29}$$

Noise correlation matrix

$$\mathbf{C}_Z = \overline{\mathbf{v}_N \mathbf{v}_N^+} \tag{4.30}$$

Noise correlation matrix transformations

$$\mathbf{C}_Z = \mathbf{Z}\ \mathbf{C}_Y\ \mathbf{Z}^+ \tag{4.31}$$

$$\mathbf{C}_Z = (\mathbf{I} + \mathbf{Z})\ \mathbf{C}_S\ (\mathbf{I} + \mathbf{Z})^+ \tag{4.32}$$

Admittance Matrix Representation

Circuit equation

$$\mathbf{i} = \mathbf{Y}\,\mathbf{v} + \mathbf{i}_N \tag{4.33}$$

Noise correlation matrix

$$\mathbf{C}_Y = \overline{\mathbf{i}_N\,\mathbf{i}_N^+} \tag{4.34}$$

Noise correlation matrix transformations

$$\mathbf{C}_Y = \mathbf{Y}\,\mathbf{C}_Z\,\mathbf{Y}^+ \tag{4.35}$$

$$\mathbf{C}_Y = (\mathbf{I} + \mathbf{Y})\,\mathbf{C}_S\,(\mathbf{I} + \mathbf{Y})^+ \tag{4.36}$$

Scattering Matrix Representation

Circuit equation

$$\mathbf{b} = \mathbf{S}\,\mathbf{a} + \mathbf{c} \tag{4.37}$$

Noise correlation matrix

$$\mathbf{C}_S = \overline{\mathbf{c}\,\mathbf{c}^+} \tag{4.38}$$

Noise correlation matrix transformations

$$\mathbf{C}_S = 1/4\,(\mathbf{I} - \mathbf{S})\,\mathbf{C}_Z\,(\mathbf{I} - \mathbf{S})^+ \tag{4.39}$$

$$\mathbf{C}_S = 1/4\,(\mathbf{I} + \mathbf{S})\,\mathbf{C}_Y\,(\mathbf{I} + \mathbf{S})^+ \tag{4.40}$$

In these families of relations **I** is the identity matrix and the impedance matrix **Z** and admittance matrix **Y** are assumed to be normalized with respect to the reference impedance of the scattering matrix **S.**

4.3 TWO-PORT-TO-THREE-PORT NOISE WAVE TRANSFORMATION

The two-port scattering matrix parameters of a three-terminal device, such as BJT, MESFET, and HEMT transistors, with one terminal grounded can be converted to three-port S-parameters when the grounded terminal is used as a third terminal. The three terminals are all referred to a common ground. This has been shown in Section 2.5 of this book. A complementary noise wave transformation must be known for evaluation of noise wave correlation matrices for transistor equivalent circuits of arbitrary topology.

Figure 4.5(a) shows a noisy three-terminal transistor considered as a three-port element. It is characterized by the matrix form

$$\begin{bmatrix} b_1' \\ b_2' \\ b_3' \end{bmatrix} = \begin{bmatrix} S_{11}' & S_{12}' & S_{13}' \\ S_{21}' & S_{22}' & S_{23}' \\ S_{31}' & S_{32}' & S_{33}' \end{bmatrix} \begin{bmatrix} a_1' \\ a_2' \\ a_3' \end{bmatrix} + \begin{bmatrix} c_1' \\ c_2' \\ c_3' \end{bmatrix} \tag{4.41}$$

where scattering parameters S' are the three-port scattering parameters.

The noise wave parameters of a two-port obtained by grounding terminal three of the three-port are obtained from the following considerations. With port-three short-circuited ($\Gamma_3 = -1$) noise waves b_3 and c_3 outgoing from port three are reflected back as ingoing noise wave $-b_3$ (see Figure 4.5(b)) [10]

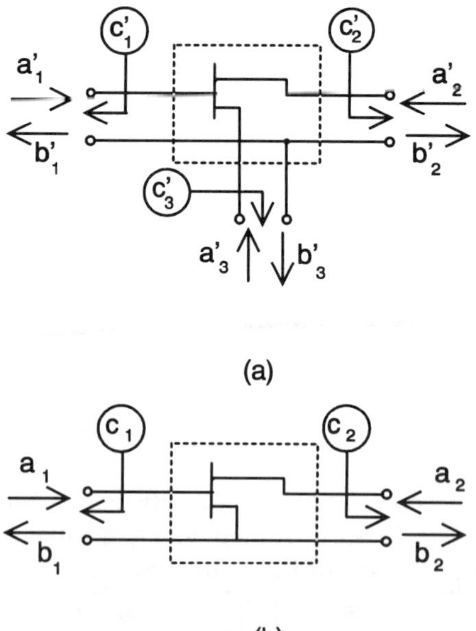

Figure 4.5 (a) Three-terminal noisy transistor considered as a three-port element with three noise wave sources at each port. (b) A two-port obtained by grounding terminal 3. Two noise wave sources represent noise generated in the element.

$$-a'_3 = b'_3 + c'_3 \tag{4.42}$$

The new two-port network is characterized by the set of equations

$$b_1 = \left(S'_{11} - \frac{S'_{13}S'_{31}}{1 + S'_{33}} \right) a_1 + \left(S'_{12} - \frac{S'_{13}S'_{32}}{1 + S'_{33}} \right) a_2 + c_1 \tag{4.43}$$

$$b_2 = \left(S'_{21} - \frac{S'_{23}S'_{31}}{1 + S'_{33}} \right) a_1 + \left(S'_{12} - \frac{S'_{23}S'_{32}}{1 + S'_{33}} \right) a_2 + c_2 \tag{4.44}$$

The quantities in the parentheses give the two-port scattering parameters, whereas the noise waves c_1 and c_2 are

$$c_1 = c'_1 - \frac{S_{13}}{1 + S_{33}} c'_3 \tag{4.45}$$

$$c_2 = c'_2 - \frac{S_{23}}{1 + S_{33}} c'_3 \tag{4.46}$$

The three-port noise waves are found by using the principle that

$$\sum_{i=1}^{3} \overline{c'_i c'^*_j} = \sum_{j=1}^{3} \overline{c'_i c'^*_j} = 0 \tag{4.47}$$

Since the noise waves in (4.46) are time averaged, we can assume that

$$c'_1 + c'_2 + c'_3 = 0 \tag{4.48}$$

From (4.44), (4.45), and (4.47), the three-port noise waves are found to be

$$c'_1 = \frac{2 + S'_{13}}{2} c_1 + \frac{S'_{13}}{2} c_2 \tag{4.49}$$

$$c'_2 = \frac{S'_{13}}{2} c_1 + \frac{2 + S'_{13}}{2} c_2 \tag{4.50}$$

$$c'_3 = -\frac{1 + S'_{33}}{2} (c_1 + c_2) \tag{4.51}$$

The three-port noise wave correlation matrix $\mathbf{C}_s'$ can be computed from the two-port noise wave correlation matrix $\mathbf{C}_s$ using the relation

$$\mathbf{C}_S' = \frac{1}{4}\mathbf{K}\,\mathbf{C}_s\,\mathbf{K}^+ \tag{4.52}$$

where

$$\mathbf{K} = \begin{bmatrix} 2 - S_{13}' & -S_{13}' \\ -S_{23}' & 2 - S_{23}' \\ 1 + S_{33}' & 1 + S_{33}' \end{bmatrix} \tag{4.53}$$

In (4.53) S_{ij}', $i, j = 1, 2, 3$, are the three-port scattering parameters that can be computed using the equations

$$S_{33}' = \frac{\sum_{i=1}^{2}\sum_{j=1}^{2} S_{ij}}{4 - \sum_{i=1}^{2}\sum_{j=1}^{2} S_{ij}} \tag{4.54}$$

$$S_{i3}' = \frac{1 + S_{33}'}{2}\left(1 - \sum_{j=1}^{2} S_{ij}\right) \qquad i = 1, 2 \tag{4.55}$$

$$S_{3j}' = \frac{1 + S_{33}'}{2}\left(1 - \sum_{i=1}^{2} S_{ij}\right) \qquad j = 1, 2 \tag{4.56}$$

$$S_{ij}' = S_{ij} + \frac{S_{i3}\,S_{3j}}{1 + S_{33}} \qquad i, j = 1, 2 \tag{4.57}$$

The inverse relation that transforms the three-port noise wave correlation matrix $\mathbf{C}'_S$ of the indefinite three-port to the two-port noise wave correlation matrix $\mathbf{C}_S$ can easily be derived using (4.44) and (4.45) as

$$\mathbf{C}_S = \mathbf{K}'\,\mathbf{C}'_S\,\mathbf{K}'^+ \tag{4.58}$$

where

$$\mathbf{K} = \begin{bmatrix} 1 & 0 & \dfrac{\Gamma_3 S_{13}}{1 - \Gamma_3 S_{33}} \\ 1 & 0 & \dfrac{\Gamma_3 S_{13}}{1 - \Gamma_3 S_{33}} \end{bmatrix} \tag{4.59}$$

4.4 NOISE WAVE CORRELATION MATRICES OF EMBEDDED MULTIPORT NETWORKS

Realistic noise modeling of microwave diodes and transistors requires the consideration of parasitic elements embedding an intrinsic chip of the device. The embedding circuit composed of parasitic stray capacitances, lead resistances, and inductances is typically passive. It produces only thermal noise. The mechanism of noise generation of the intrinsic device is of a more complicated nature, and it is very convenient to consider and model it separately. The effects of a parasitic element circuit on the correlation matrix of the entire device may be computed using appropriate embedding formulas.

Figure 4.6 presents a noisy multiport subnetwork with scattering matrix **S** embedded in a noisy subnetwork with scattering matrix **T.** It is assumed that noise properties of both subnetworks are described by noise wave correlation matrices $\mathbf{C}_S$ and $\mathbf{C}_T$, respectively. The noise wave correlation matrix of the resultant network is computed by partitioning ports of the embedding subnetwork into two categories: external and internal ports. Accordingly, the system of equations

$$\mathbf{b} = \mathbf{T}\ \mathbf{a} + \mathbf{c} \tag{4.60}$$

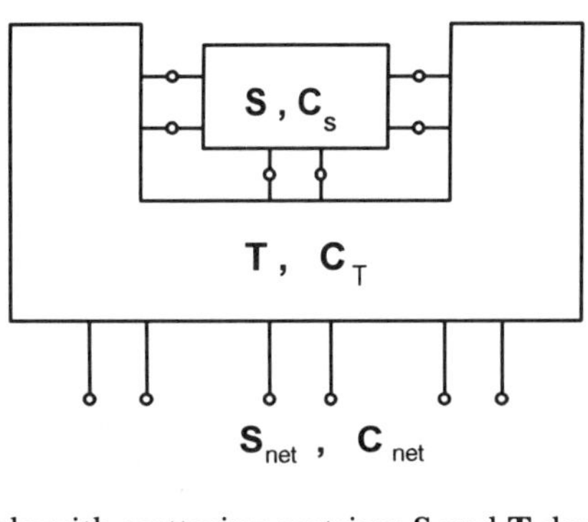

Figure 4.6 Two subnetworks with scattering matrices **S** and **T** described by their noise wave correlation matrices $\mathbf{C}_S$ and $\mathbf{C}_T$ and connected by internal ports. The resultant scattering and noise wave correlation matrices are $\mathbf{S}_{net}$ and $\mathbf{C}_{net}$.

can be written as [11]

$$\begin{bmatrix} \mathbf{b}_e \\ \mathbf{b}_i \end{bmatrix} = \begin{bmatrix} \mathbf{T}_{ee} & \mathbf{T}_{ei} \\ \mathbf{T}_{ie} & \mathbf{T}_{ii} \end{bmatrix} \begin{bmatrix} \mathbf{a}_e \\ \mathbf{a}_i \end{bmatrix} + \begin{bmatrix} \mathbf{c}_e \\ \mathbf{c}_i \end{bmatrix} \tag{4.61}$$

where $\mathbf{a}_e$, $\mathbf{b}_e$, $\mathbf{c}_e$ and $\mathbf{a}_i$, $\mathbf{b}_i$, $\mathbf{c}_i$ are vectors of noise waves, respectively, at the external and internal ports of the subnetwork **T.**

The embedded subnetwork is described by the system of equations

$$\mathbf{b}_S = \mathbf{S}\, \mathbf{a}_S + \mathbf{c}_S \tag{4.62}$$

The connection of internal ports of the embedding subnetwork **T** with appropriate ports of the subnetwork **S** impose restrictions on the vectors of ingoing and outgoing noise waves in the forms

$$\mathbf{a}_i = \mathbf{b}_S \tag{4.63}$$

$$\mathbf{b}_i = \mathbf{a}_S \tag{4.64}$$

From (4.62), (4.63), (4.64), and the second equation of (4.61), by first eliminating $\mathbf{b}_i$, we obtain

$$\mathbf{a}_i = [\mathbf{S}^{-1} - \mathbf{T}_{ii}]^{-1}\, \mathbf{T}_{ie}\, \mathbf{a}_e + [\mathbf{I} - \mathbf{S}\,\mathbf{T}_{ii}]^{-1}\, \mathbf{c}_S + [\mathbf{S}^{-1} - \mathbf{T}_{ii}]^{-1}\mathbf{c}_i \tag{4.65}$$

and, after eliminating $\mathbf{a}_i$ from the first equation of (4.61)

$$\begin{aligned} \mathbf{b}_e = \{&\mathbf{T}_{ee} + \mathbf{T}_{ei}\, [\mathbf{S}^{-1} - \mathbf{T}_{ii}]^{-1}\, \mathbf{T}_{ie}\}\, \mathbf{a}_e \\ &+ \mathbf{T}_{ei}\, [\mathbf{I} - \mathbf{S}\,\mathbf{T}_{ii}]^{-1}\, \mathbf{c}_S + \mathbf{T}_{ei}\, [\mathbf{I} - \mathbf{S}\,\mathbf{T}_{ii}]^{-1}\, \mathbf{S}\, \mathbf{c}_i + \mathbf{c}_e \end{aligned} \tag{4.66}$$

In (4.66)

$$\mathbf{c}_{out} = \mathbf{T}_{ei}\, [\mathbf{I} - \mathbf{S}\,\mathbf{T}_{ii}]^{-1}\, \mathbf{c}_S + \mathbf{T}_{ei}\, [\mathbf{I} - \mathbf{S}\,\mathbf{T}_{ii}]^{-1}\, \mathbf{S}\, \mathbf{c}_i + \mathbf{c}_e \tag{4.67}$$

is a vector of resultant noise waves of the whole network referred to the external ports.

Using (4.67) the resultant noise wave correlation matrix of the network can be found as [11]

$$\mathbf{C}_{net} = \Lambda\, \mathbf{C}_S\, \Lambda^+ + [\mathbf{I} \mid \Lambda\, \mathbf{S}]\, \mathbf{C}_T\, [\mathbf{I} \mid \Lambda\, \mathbf{S}]^+ \tag{4.68}$$

where

$$\mathbf{C}_T = \begin{bmatrix} \overline{\mathbf{c}_e\mathbf{c}_e^+} & \overline{\mathbf{c}_e\mathbf{c}_i^+} \\ \overline{\mathbf{c}_i\mathbf{c}_e^+} & \overline{\mathbf{c}_i\mathbf{c}_i^+} \end{bmatrix} \tag{4.69}$$

is the noise wave correlation matrix of network **T**, Λ is the matrix given as

$$\Lambda = \mathbf{T}_{ei}\,[\mathbf{I} - \mathbf{S}\,\mathbf{T}_{ii}]^{-1} \tag{4.70}$$

and **I** is the identity matrix.

4.5 NOISE MODELING OF MICROWAVE NETWORK ELEMENTS

To perform noise analysis of a microwave network it is necessary to represent each network element by its scattering matrix and by its noise wave correlation matrix.

4.5.1 Noise Wave Correlation Matrices of Passive Multiports

In general, elements of microwave circuits can be divided into two groups: passive and active multiports. Lossy passive multiports generate only thermal noise. The noise wave correlation matrix of a lossy passive multiport is derived directly from its scattering matrix [8, 12] as

$$\mathbf{C}_S = kT\,(\mathbf{I} + \mathbf{S}\,\mathbf{S}^+) \tag{4.71}$$

where k denotes Boltzmann's constant,

T the physical absolute temperature of the multiport, and

I is an identity matrix.

Of course, the above relation is applicable also to lossy passive two-ports.

4.5.2 Noise Wave Correlation Matrices of Active Two-Ports

For active two-ports, as for example BJT, MESFET, and HEMT transistors, their correlation matrices can be obtained from their noise equivalent circuits [13, 19].

Example 4.1

In the last few years, Pospieszalski's approach to noise modeling of FET (HEMT) has gained much attention and popularity in the microwave community [19]. In his model presented in Figure 4.7(a) noise properties of an intrinsic chip are

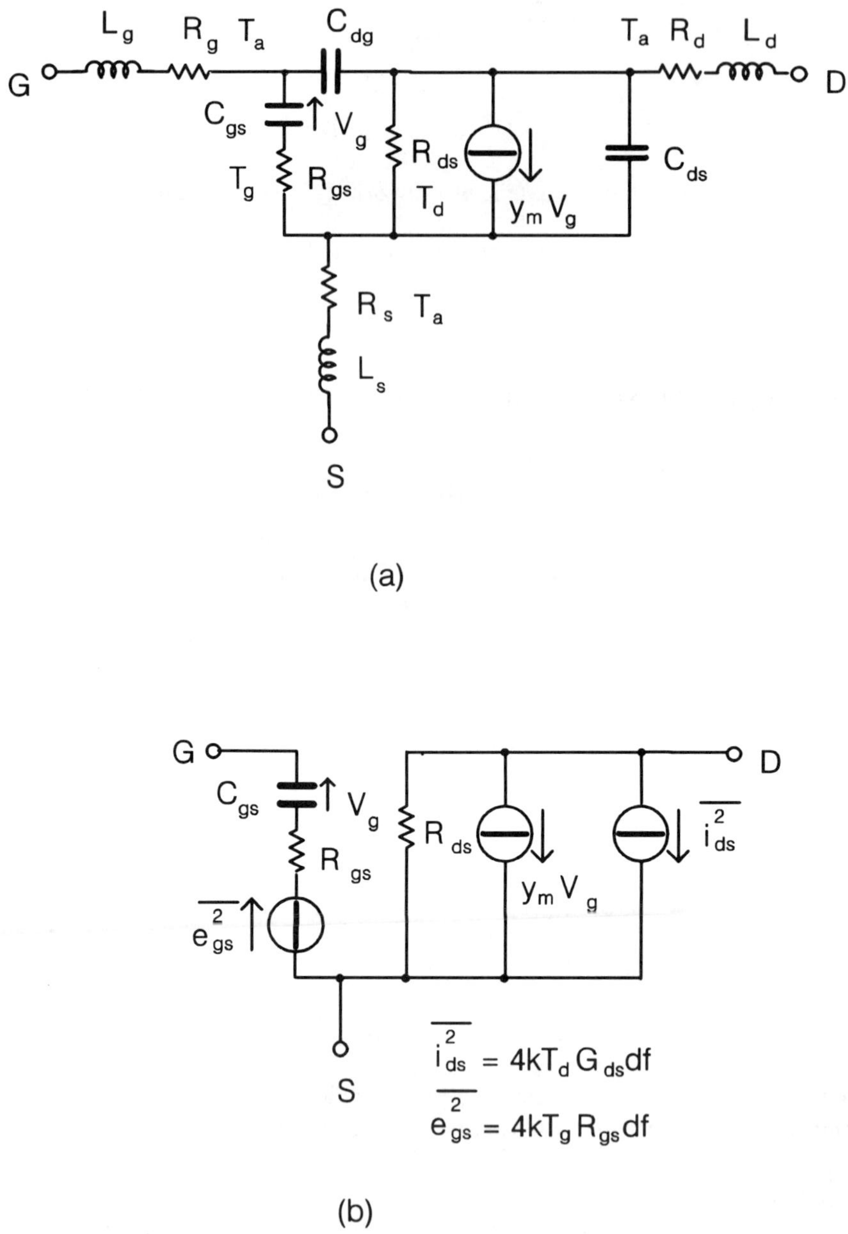

Figure 4.7 (a) Equivalent circuit of FET (HEMT) chip. Noise properties of an intrinsic chip are represented by equivalent temperatures: T_g of R_{gs} and T_d of R_{ds}. Noise contributions of ohmic resistances R_s, R_d, and R_g are determined by physical temperature T_a of a chip. (b) Noise equivalent circuit of an intrinsic chip.

represented by the equivalent temperature T_g of the gate-source resistance R_{gs} and by the equivalent temperature T_d of the drain-source resistance R_{ds}.

Noise contributions of ohmic resistances R_g, R_d, and R_s are determined by physical temperature T_a of a chip. Considering the noise equivalent circuit of an intrinsic chip given in Figure 4.7(b) (common source configuration) it is very easy to derive equations determining noise waves c_1 and c_2 emanating, respectively, from the gate port and from the drain port.

Using a circuit presented in Figure 4.8 we derive noise currents and noise voltages at input and output port terminals loaded with impedances equal to reference impedance Z_R.

For the input port,

$$I_1 = \frac{e_{gs}}{R_{gs} + Z_R + 1/j\,\omega\,C_{gs}} \tag{4.72}$$

$$V_1 = \frac{e_{gs}\,Z_R}{R_{gs} + Z_R + 1/j\,\omega\,C_{gs}} \tag{4.73}$$

For the output port,

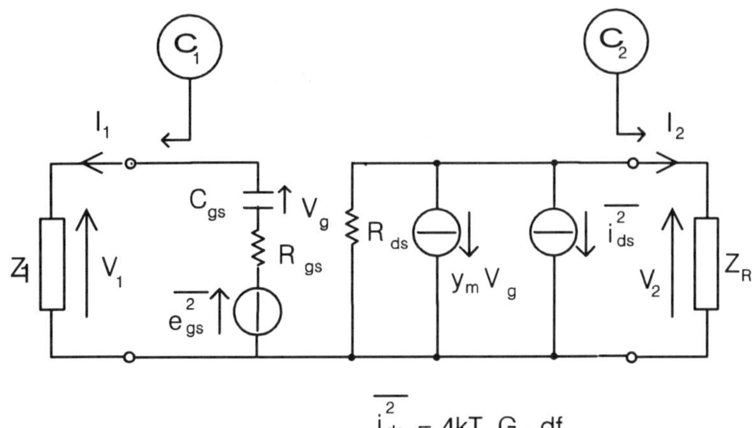

$$\overline{i_{ds}^2} = 4kT_d\,G_{ds}df$$

$$\overline{e_{gs}^2} = 4kT_g\,R_{gs}df$$

Figure 4.8 Noise equivalent circuit of an intrinsic chip loaded at the input port and at the output port by reference impedances Z_R.

$$I_2 = -\frac{R_{ds}}{R_{ds} + Z_r}(i_{ds} + g_m V_g) \tag{4.74}$$

$$V_2 = -\frac{R_{ds} Z_R}{R_{ds} + Z_r}(i_{ds} + g_m V_g) \tag{4.75}$$

where

$$V_g = \frac{e_{gs} j \omega C_{gs}}{R_{gs} + Z_R + 1/j \omega C_{gs}} \tag{4.76}$$

Using the definition of ingoing noise wave given by (4.2) and (4.72) to (4.76) we have

$$c_1 = \frac{V_1 + Z_R I_1}{2\sqrt{Z_R}} = \frac{\sqrt{Z_R}}{R_{gs} + Z_R + 1/j \omega C_{gs}} e_{gs} \tag{4.77}$$

$$c_2 = \frac{V_2 + Z_R I_2}{2\sqrt{Z_R}} = -\frac{R_{ds}\sqrt{Z_R}}{R_{gs} + Z_R}(i_{ds} + g_m V_g) \tag{4.78}$$

The input port noise wave c_1 is due only to the thermal noise of input resistance R_{gs} at temperature T_g, while the output noise wave c_2 results from the correlated noise from the gate (thermal noise generated by R_{gs} induces a voltage V_g across C_{gs} that is transferred to the drain by transconductance g_m) and the noise generated by output resistance R_{ds} at equivalent temperature T_d.

The noise wave parameters of the intrinsic chip are given by

$$\overline{|c_1|^2} = \overline{c_1 c_1^*} = \frac{4 k T_g R_{gs} Z_R \omega^2 C_{gs}^2}{1 + \omega^2 C_{gs}^2 (R_{gs} + Z_R)^2} \tag{4.79}$$

$$\overline{c_1 c_2^*} = \frac{-j 4 k T_g R_{ds} R_{gs} Z_R g_m \omega C_{gs}}{(R_{ds} + Z_R)^2 [1 + \omega^2 C_{gs}^2 (R_{gs} + Z_R)^2]} \tag{4.80}$$

$$\overline{|c_2|^2} = \frac{4 k T_g R_{ds} R_{gs} Z_R g_m^2}{(R_{ds} + Z_R)^2 [1 + \omega^2 c_{gs}^2 (R_{gs} + Z_R)^2]} + \frac{4 k T_d R_{ds} Z_R}{(R_{ds} + Z_R)^2} \tag{4.81}$$

In (4.79) to (4.81) it is assumed that $\Delta f = 1$ Hz and that $\overline{|e_{gs}|^2} = 4 k T_g \Delta f R_{gs}$

and $\overline{|i_{ds}|^2} = 4\,k\,T_d\,\Delta f\,1/R_{ds}$. The noise voltage source $\overline{|e_{gs}|^2}$ models a noise process in drain and gate, whereas the current noise source $|i_{ds}|^2$ models a noise process that produces the noise current only in a drain circuit [19].

The procedure of derivation of the noise wave correlation matrix from noise equivalent circuit of active device presented here is a general one. It may be used for any noise model of BJTs, MESFETs, or HEMTs found in literature [13, 14, 17, 18].

In cases when the correlation matrix cannot be derived from theoretical investigations, measurements of noise characteristics provide the required information. Proper measurement techniques are used to obtain four device noise parameters: F_{min}, minimum noise figure; R_n, the equivalent noise resistance (as defined in (4.14)); and $Y_{opt} = G_{opt} + j\,B_{opt}$, the optimum signal generator admittance at which the noise figure has a minimum value [20–23]. The other equivalent four noise parameters are: F_{min} and g_n, the equivalent noise conductance; and $Z_{opt} = R_{opt} + j\,X_{opt}$, the optimum signal generator impedance, or F_{min}, R_n, and Γ_{opt}, the optimum reflection coefficient of the signal generator port. With these parameters estimated, the noise wave correlation matrix of the active two-port is obtained as

$$\begin{aligned} C_{S11} = k\,T_0 \frac{R_n}{Z_R} &(|S_{11}|^2 - 1 - 2\,\mathrm{Re}\{S_{11}\}) \\ &+ (F_{emin} - 2\,R_n\,Y_{opt})\,(|S_{11}|^2 - 1) + 4\,R_n\,B_{opt}\,\mathrm{Im}\{S_{11}\} \\ &+ Z_R\,R_n\,|Y_{opt}|^2\,(|S_{11}|^2 + 1 + 2\,\mathrm{Re}\{S_{11}\}) \end{aligned} \tag{4.82}$$

$$\begin{aligned} C_{S12} = C^*_{S21} = k\,T_0 \frac{S^*_{21}}{Z_R} &(S_{11} - 1)\,(R_n - Z_R^2\,R_n\,|Y_{opt}|^2) \\ &+ S_{11}\,S^*_{21}\,(F_{emin} - 2\,R_n\,G_{opt}) - j\,2\,S^*_{21}\,R_n\,B_{opt} \end{aligned} \tag{4.83}$$

$$C_{S22} = k\,T_0 \frac{|S_{21}|^2}{Z_R}\,[R_n + Z_R\,(F_{emin} - 2\,R_n\,G_{opt}) + Z_R^2\,R_n\,|Y_{opt}|^2] \tag{4.84}$$

In the above representations $F_{emin} = F_{min} - 1$ is the minimum excess noise figure, R_n is the equivalent noise resistance, $Y_{opt} = G_{opt} + j\,B_{opt}$ is the optimum signal generator internal admittance at which $F = F_{min}$, and Z_R is the reference impedance of the scattering parameters. Often, instead of R_n, the noise parameter N is used. N is defined as

$$N = 4\,R_n\,G_{opt} \tag{4.85}$$

It is known that F_{min} and the parameter N are invariant under transformation through lossless reciprocal two-ports connected to the input of a noisy two-port [20]. It has also been observed that in order for F_{min} and N to represent a physical two-port, the following inequality must be satisfied [21]:

$$F_{emin} \leq N \tag{4.86}$$

This inequality follows directly from the property of the correlation matrices, which must be Hermitian and nonnegative definite.

Instead of the optimum signal generator admittance Y_{opt}, the optimum reflection coefficient of the signal generator port is often used. It is equal to

$$\Gamma_{opt} = \frac{Y_R - Y_{opt}}{Y_R + Y_{opt}} \tag{4.87}$$

where Y_R is the reference admittance of the generator port.

In terms of F_{emin}, Γ_{opt}, and R_n the noise wave parameters are

$$\overline{c_{N1}|^2} = k\, T_0\, F_{emin}\, (|s_{11}|^2 - 1) + \frac{4\, k\, T_0\, R_n\, |1 - s_{11}\, \Gamma_{opt}|^2}{Z_R\, |1 + \Gamma_{opt}|^2} \tag{4.88}$$

$$\overline{|c_{N2}|^2} = |s_{21}|^2 \left(K\, T_0\, F_{emin} + \frac{4\, k\, T_0\, R_n\, |\Gamma_{opt}|^2}{Z_R\, |1 + \Gamma_{opt}|^2} \right) \tag{4.89}$$

$$\overline{c_{N1}\, c_{N2}{}^*} = \frac{-4\, k\, T_0\, R_n\, s_{21}^*\, \Gamma_{opt}}{Z_R\, |1 + \Gamma_{opt}|^2} + \frac{s_{11}}{s_{21}} \overline{|c_{N2}|^2} \tag{4.90}$$

where Z_R is the reference impedance.

It is also possible to directly measure noise wave parameters of microwave active devices using noise wave correlation analyzers or interferometer measurement systems. These measurement techniques will be discussed in Chapter 9.

4.6 BASIC RELATIONSHIPS FOR NOISY TWO-PORTS

A very important parameter and aspect of microwave network design is the noise figure. The noise figure is defined as

$$F = 1 + \frac{P_{N\ int}}{P_{N\ G}} = 1 + F_e \tag{4.91}$$

where

$P_{N\ int}$ denotes the (available) noise power at the output port of the circuit originated from noise sources acting within the circuit, and

$P_{N\ G}$ denotes the (available) noise power at the output port of the circuit originated from the equivalent thermal noise source of the input port termination (signal generator impedance).

In the definition of the noise figure it is assumed that the load is noisefree. According to the IEEE definition of the noise figure F the (available) noise power of the signal generator must be equal to the thermal noise power of the generator impedance in standard temperature $T_0 = 290$ K.

The noise figure of a system may also be defined as the ratio of signal-to-noise ratios existing at input and output

$$F = \frac{(P_S/P_N)_{in}}{(P_S/P_N)_{out}} = \frac{(P_S)_{in}/P_{NG}}{G_S(P_S)_{in}/G_N\,(P_{NG} + P_{N\ int})} \tag{4.92}$$

If the power gains for signal and noise are the same, that is, $G_S = G_N = G$, and the input noise power corresponds to the signal generator impedance, being in temperature $T = 290$ K, both definitions are equivalent.

When the two-port circuit is driven by a signal generator with the internal admittance $Y_G = G_G + jB_G$, its noise figure is written in the form [2]

$$F = F_{min} + \frac{R_n}{G_G}\,|Y_G - Y_{opt}|^2 \tag{4.93}$$

The other equations equivalent to (4.89) are

$$F = F_{min} + \frac{g_n}{R_G}\,|Z_G - Z_{opt}|^2 \tag{4.94}$$

where $Z = R_G + jX_G$ is the internal impedance of the signal generator, or

$$F = F_{min} + N\frac{|\Gamma_G - \Gamma_{opt}|^2}{(1 - |\Gamma_G|^2)\,(1 - |\Gamma_{opt}|)^2} \qquad N = 4\,R_n\,G_{opt} \tag{4.95}$$

and

$$F = F_{min} + 4\,\frac{R_n}{Z_R}\,\frac{|\Gamma_G - \Gamma_{opt}|^2}{|1 + \Gamma_{opt}|^2\,(1 - |\Gamma_G|^2)} \tag{4.96}$$

where Γ_G is the reflection coefficient of the signal generator port with respect to Z_R and Z_R is the real reference impedance of the input port of the two-port.

According to (4.29), (4.30), and (4.31) it is obvious that the device noise parameters and the noise figure of a two-port can be expressed as functions of the entries of the correlation matrices, as it is easily predictable from a physical point of view. These relations are

$$R_N = C_{A11} \tag{4.97}$$

$$Y_{opt} = G_{opt} + j\,B_{opt} = \left[\frac{C_{A22}}{C_{A11}} - \left(\mathrm{Im}\left\{\frac{C_{A12}}{C_{A11}}\right\}\right)^2\right]^{1/2} + j\,\mathrm{Im}\left\{\frac{C_{A12}}{C_{A11}}\right\} \tag{4.98}$$

$$F_{min} = 1 + 2(\mathrm{Re}\{C_{A12}\} + C_{A11}G_{opt}) \tag{4.99}$$

Instead of the optimum signal generator admittance Y_{opt}, the optimum reflection coefficient of the signal generator port may be used (see (4.87)).

In (4.97), (4.98), and (4.99), C_{A11}, C_{A12}, and C_{A22} are the noise correlation matrix elements for the chain matrix representation defined by (4.23).

Standard noise parameters of an active two-port may also be derived from noise wave parameters. These relations are [11]

$$R_n = \frac{Z_R}{4\,kT_0}\overline{\left| c_1 - c_2\left(\frac{1+s_{11}}{s_{21}}\right)\right|^2}$$

$$= \frac{Z_R}{4\,kT_0}\left[C_{S11} - 2\,\mathrm{Re}\left\{C_{S12}\frac{1+s_{11}}{s_{21}}\right\} + C_{S22}\left|\frac{1+s_{11}}{s_{21}}\right|^2\right] \tag{4.100}$$

$$kT_0F_{emin} = \frac{\overline{|c_2|^2} - \overline{|c_1s_{21} - c_2s_{11}|^2}|\Gamma_{opt}|^2}{|s_{21}|^2(1+|\Gamma_{opt}|^2)}$$

$$= \frac{C_{S22} - (C_{S11}|s_{21}|^2 - 2\,\mathrm{Re}\{//C_{S12}s_{21}s_{11}^*\} + C_{S22}|s_{11}|^2)|\Gamma_{opt}|^2}{|s_{21}|^2(1+|\Gamma_{opt}|^2)} \tag{4.101}$$

$$\Gamma_{opt} = \frac{\eta}{2}\left(1 - \sqrt{1 - \frac{4}{|\eta|^2}}\right) \tag{4.102}$$

where

$$\eta = \frac{\overline{|c_2|^2} + \overline{|c_1 s_{21} - c_2 s_{11}|^2}}{\overline{|c_2|^2}\, s_{11} - \overline{c_1 c_2^*}\, s_{21}}$$

$$= \frac{C_{S22} + C_{S11}\,|s_{21}|^2 - 2\,\mathrm{Re}\{C_{S12}\, s_{21}\, s_{11}^*\} + C_{S22}\,|s_{11}|^2}{C_{S22}\, s_{11} - C_{S12}\, s_{21}} \tag{4.103}$$

To derive the above equations the following identity was used

$$\overline{|c_1\alpha - c_2\beta|^2} = \overline{|c_1|^2}\,|\alpha|^2 - 2\,\mathrm{Re}\{\overline{c_1 c_2^*}\,\alpha\,\beta^*\} + \overline{|c_2|^2}\,|\beta|^2 \tag{4.104}$$

The noise figure expressed by the noise wave matrix elements is given by

$$F = 1 + \frac{C_{S11}\left|\dfrac{S_{21}\Gamma_G}{1-S_{11}\Gamma_G}\right|^2 + C_{S22} + 2\,\mathrm{Re}\left\{C_{S12}\dfrac{S_{21}\Gamma_G}{1-S_{11}\Gamma_G}\right\}}{(1-|\Gamma_G|^2)\left|\dfrac{S_{21}}{1-S_{11}\Gamma_G}\right|^2} \tag{4.105}$$

The above equation may be written in a more compact form as

$$F = 1 + \frac{\boldsymbol{\alpha}^T\,\mathbf{C}_s\,\boldsymbol{\alpha}}{k\,T_0\,(1-|\Gamma_G|^2)} \tag{4.106}$$

where α is the vector

$$\alpha = \begin{bmatrix} \Gamma_S \\ \hdashline \dfrac{1-\Gamma_S\, s_{11}}{s_{21}} \end{bmatrix} \tag{4.107}$$

Another very important noise parameter is noise measure M, which is defined as [7]

$$M = \frac{F-1}{1-1/G_A} \tag{4.108}$$

and may be written in terms of noise wave parameters as

$$k\,T_0\,M = \frac{-\boldsymbol{\alpha}^T \mathbf{C}_s\,\boldsymbol{\alpha}}{\boldsymbol{\alpha}^T(\mathbf{I} - \mathbf{S}\mathbf{S}^+)\,\boldsymbol{\alpha}} \tag{4.109}$$

where $\mathbf{S}$ and $\mathbf{C}_s$ are the scattering matrix and noise wave correlation matrix of the two-port, respectively. As always in this book, we have assumed a 1-Hz bandwidth.

When a zero value of the signal source reflection coefficient is assumed, (4.106) and (4.109) are simplified to

$$F_{emin} = \frac{\overline{|c_2|^2}}{k\,T_0\,|s_{21}|^2} \tag{4.110}$$

and

$$k\,T_0\,M = \frac{\overline{|c_2|^2}}{|s_{21}|^2 + |s_{22}|^2 - 1} \tag{4.111}$$

4.7 NOISE ANALYSIS OF ARBITRARY TOPOLOGY LINEAR MULTIPORT NETWORKS

In the method of microwave network noise analysis to be presented here, it is assumed that each linear noisy network may be considered as the interconnection of multiports. Each linear element in the network is represented by its noiseless equivalent having the same scattering matrix $\mathbf{S}$ as the original network. As is shown in Figure 4.3, noise generated in each element is represented by mutually correlated noise wave sources, one source at each port.

In matrix notation, a set of linear equations that relates complex amplitudes of noise waves at ports of a circuit element has a form [24, 25]

$$\mathbf{b}^{(k)} = \mathbf{S}^{(k)}\,\mathbf{a}^{(k)} + \mathbf{c}^{(k)} \tag{4.112}$$

where $\mathbf{S}^{(k)}$ is the scattering matrix of the kth element, $\mathbf{a}^{(k)}$ and $\mathbf{b}^{(k)}$ are vectors of ingoing and outgoing noise waves at its ports, and $\mathbf{c}^{(k)}$ is a vector of mutually correlated noise wave sources that represent noise generated in the element. The noise waves from these sources radiate out of the ports and do not depend on incident noise waves $\mathbf{a}^{(k)}$.

Figure 4.9 presents a general circuit composed of m elements (multiports) connected together by their ports.

Considering all m elements of the circuit, we have a set of linear equations whose matrix form is [22]

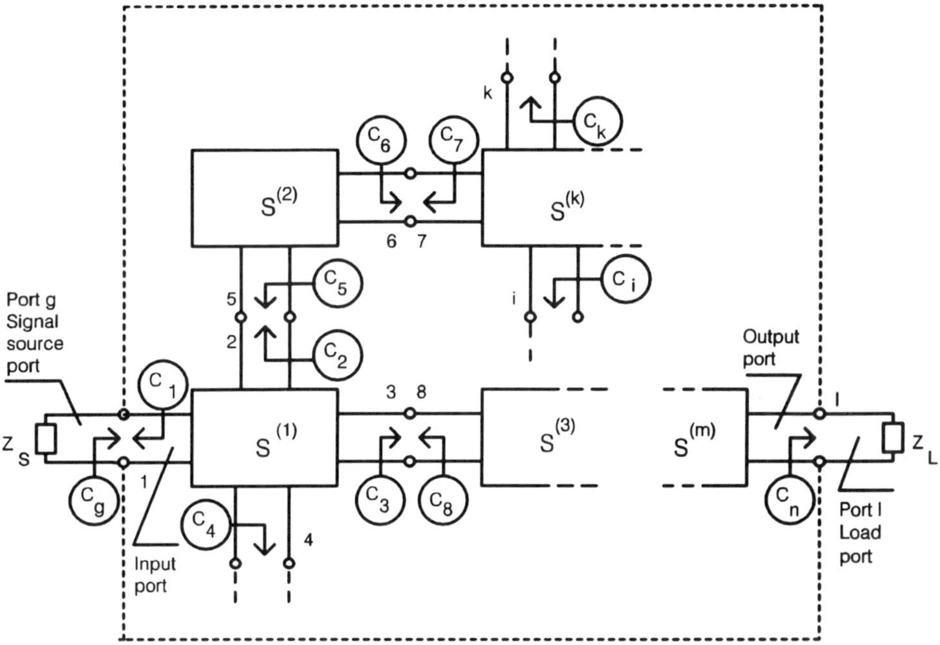

Figure 4.9 Equivalent circuit of a multiport network with noiseless elements and noise wave sources at each port. Each element of the network is described by its scattering matrix **S** and by its noise wave correlation matrix $\mathbf{C}_s$.

$$\mathbf{b} = \mathbf{S}\,\mathbf{a} + \mathbf{c} \tag{4.113}$$

where

$$\mathbf{a} = \begin{bmatrix} \mathbf{a}^{(1)} \\ \mathbf{a}^{(2)} \\ \vdots \\ \mathbf{a}^{(k)} \\ \vdots \\ \mathbf{a}^{(m)} \end{bmatrix} \qquad \mathbf{b} = \begin{bmatrix} \mathbf{b}^{(1)} \\ \mathbf{b}^{(2)} \\ \vdots \\ \mathbf{b}^{(k)} \\ \vdots \\ \mathbf{b}^{(m)} \end{bmatrix} \qquad \mathbf{c} = \begin{bmatrix} \mathbf{c}^{(1)} \\ \mathbf{c}^{(2)} \\ \vdots \\ \mathbf{c}^{(k)} \\ \vdots \\ \mathbf{c}^{(m)} \end{bmatrix} \tag{4.114}$$

and

$$\mathbf{S} = \begin{bmatrix} \mathbf{S}^{(1)} & \mathbf{0} & \dots & \mathbf{0} & \dots & \mathbf{0} \\ \mathbf{0} & \mathbf{S}^{(2)} & & & & \mathbf{0} \\ \vdots & & \ddots & & & \vdots \\ \mathbf{0} & & & \mathbf{S}^{(k)} & & \mathbf{0} \\ \vdots & & & & \ddots & \mathbf{0} \\ \mathbf{0} & \dots & \dots & \mathbf{0} & \mathbf{0} & \mathbf{S}^{(m)} \end{bmatrix} \tag{4.115}$$

The connections between the m elements impose constraints on the vectors **a** and **b,** which can be represented as a matrix equation

$$\mathbf{b} = \mathbf{\Gamma}\,\mathbf{a} \tag{4.116}$$

where $\mathbf{\Gamma}$ is the connection matrix defined in Section 3.1.

After elimination of the vector **b** from (4.113) and (4.116) we obtain

$$\mathbf{W}\,\mathbf{a} = \mathbf{c} \tag{4.117}$$

where

$$\mathbf{W} = \mathbf{\Gamma} - \mathbf{S} \tag{4.118}$$

is the connection scattering matrix of the analyzed network defined and discussed in Chapter 3.

Using (4.117) we are able to get a correlation matrix of the incident noise waves at all circuit ports. Because

$$\mathbf{a} = \mathbf{W}^{-1}\,\mathbf{c} \tag{4.119}$$

it follows that

$$\overline{\mathbf{a}\,\mathbf{a}^{+}} = \mathbf{W}^{-1}\,\overline{\mathbf{c}\,\mathbf{c}^{+}}\,(\mathbf{W}^{-1})^{+} = \mathbf{W}^{-1}\,\mathbf{C}\,(\mathbf{W}^{-1})^{+} \tag{4.120}$$

where the daggers indicate the Hermitian complex conjugate of vectors and matrices.

In (4.120)

$$\mathbf{C} = \overline{\mathbf{c}\,\mathbf{c}^{+}} \tag{4.121}$$

is the correlation matrix of the noise wave sources representing noise generated in all circuit elements.

Because the noise wave sources $\mathbf{c}^{(k)}$ of the kth element are uncorrelated with those of any other circuit element, the correlation matrix $\mathbf{C}$ is a block diagonal matrix of the form

$$\mathbf{C} = \overline{\mathbf{c}\,\mathbf{c}^{+}} = \begin{bmatrix} \mathbf{C}_{\mathrm{s}}^{(1)} & \mathbf{0} \ldots & \mathbf{0} & \ldots & \mathbf{0} \\ \mathbf{0} & \mathbf{C}_{\mathrm{s}}^{(2)} & & & \vdots \\ \vdots & & \ddots & & \vdots \\ \mathbf{0} & & \mathbf{C}_{\mathrm{s}}^{(k)} & & \mathbf{0} \\ \vdots & & & \ddots & \vdots \\ \mathbf{0} & \ldots\ldots & \mathbf{0} & \ldots & \mathbf{C}_{\mathrm{s}}^{(m)} \end{bmatrix} \tag{4.122}$$

in which $\mathbf{C}_{\mathrm{s}}^{(1)}, \mathbf{C}_{\mathrm{s}}^{(2)}, \ldots, \mathbf{C}_{\mathrm{s}}^{(m)}$ are correlation matrices of the noise wave sources of individual network elements and the $\mathbf{0}$s represent null matrices.

4.7.1 The Algorithm for Noise Figure Computation

If l is a number of the load impedance port of the analyzed circuit, then the active noise power at the output port is [24]

$$P_{\mathrm{N}} = (\overline{\mathbf{a}\,\mathbf{a}^{+}})_{ll}\,(1 - |S_{ll}|^2) = N_l\,(1 - |S_{ll}|^2) \tag{4.123}$$

where $N_l = (\overline{\mathbf{a}\,\mathbf{a}^{+}})_{ll}$ is the lth diagonal element of the correlation matrix $\overline{\mathbf{a}\,\mathbf{a}^{+}}$,

$$S_{ll} = \frac{Z_{\mathrm{L}} - Z_{\mathrm{R}}}{Z_{\mathrm{L}} + Z_{\mathrm{R}}} \tag{4.124}$$

is the reflection coefficient of the output port load,
Z_{L} denotes the load impedance, and
Z_{R} the reference impedance of the load impedance port.

The evaluation of N_l can be derived easily from (4.120). In fact, if only the lth diagonal element of the correlation matrix $\overline{\mathbf{a}\,\mathbf{a}^{+}}$ has to be determined, by letting $\mathbf{e}_l$ be a vector whose elements are all zeros except a 1 in position l; that is,

$$\mathbf{e}_l^{\mathrm{T}} = [0, \ldots, 0, 1, 0, \ldots, 0] \tag{4.125}$$

we have

$$\begin{aligned} N_l &= (\overline{\mathbf{a}\,\mathbf{a}^{+}})_{ll} = \mathbf{e}_l^{\mathrm{T}}\,\overline{\mathbf{a}\,\mathbf{a}^{+}}\,\mathbf{e}_l \\ &= \mathbf{e}_l^{+}\,\mathbf{W}^{-1}\,\mathbf{C}\,(\mathbf{W}^{+})^{-1}\,\mathbf{e}_l \\ &= [(\mathbf{W}^{+})^{-1}\,\mathbf{e}_l]^{+}\,\mathbf{C}\,(\mathbf{W}^{+})^{-1}\,\mathbf{e}_l \end{aligned} \tag{4.126}$$

Relation (4.126) may be also written in the form

$$N_l\,(\mathbf{a}\,\mathbf{a}^{+})_{ll} = \underline{\mathscr{A}}^{+}\,\mathbf{C}\,\underline{\mathscr{A}} \tag{4.127}$$

where a vector

$$\underline{\mathscr{A}} = (\mathbf{W}^{+})^{-1}\mathbf{e}_l = (\mathbf{W}^{\mathrm{T}*})^{-1}\,\mathbf{e}_l \tag{4.128}$$

is the solution vector of a system of equations

$$\mathbf{W}^{+}\,\underline{\mathscr{A}} = \mathbf{e}_l \tag{4.129}$$

whose coefficient matrix is equal to the complex transpose of the connection scattering matrix **W** of the analyzed circuit and has $\mathbf{e}_l$ as its right-hand-side vector.

Because of that noise powers P_{Nint} and P_{NG} can be computed using simple matrix multiplications described by (4.127) and (4.128).

The following steps are carried out by the numerical algorithm.

Step 1: The connection scattering matrix **W** of the analyzed circuit is built and computed.

Step 2: The noise correlation matrices of the individual passive and active circuit elements are found. The noise correlation matrix **C** defined by (4.122) is built and computed. Two diagonal elements of **C** relative to ports that belong to the signal source impedance and to the load impedance are set to zero.

$$
\begin{array}{rc}
 & \begin{array}{cccccccc} \phantom{\mathbf{C}_s^{(1)}} & \phantom{\mathbf{C}_s^{(2)}} & & g & & \phantom{\mathbf{C}_s^{(k)}} & l & \phantom{\mathbf{C}_s^{(m)}} \end{array} \\
\mathbf{C}_{\mathrm{int}} = \begin{array}{c} \\ \\ \\ g \\ \\ \\ l \\ \\ \end{array} &
\left[\begin{array}{cccccccc}
\mathbf{C}_s^{(1)} & & & \vdots & & & \vdots & \\
 & \mathbf{C}_s^{(2)} & & \vdots & & & \vdots & \\
 & & \ddots & \vdots & & & \vdots & \\
\cdots & \cdots & \cdots & \mathbf{0} & & & \vdots & \\
 & & & & \ddots & & \vdots & \\
 & & & & & \mathbf{C}_s^{(k)} & \vdots & \\
\cdots & \cdots & \cdots & \cdots & \cdots & \cdots & \mathbf{0} & \\
 & & & & & & & \mathbf{C}_s^{(m)}
\end{array}\right]
\end{array}
\tag{4.130}
$$

g = signal generator port number
l = load impedance port number

Both zeros represent no noise power generated by the signal generator and load impedances.

Step 3: Relative to the load impedance, the port diagonal element $N_{\text{int}\,l}$ of the matrix $\overline{\mathbf{a}\,\mathbf{a}^{+}}$ is computed using (4.127) as

$$N_{\text{int}\,l} = \underline{\mathcal{A}}^{+}\,\mathbf{C}_{\text{int}}\,\underline{\mathcal{A}} \tag{4.131}$$

A value of this element multiplied by $(1 - |S_{ll}|^2)$ equals P_{Nint}.

Step 4: The noise correlation matrix **C** defined by (4.122) is built and computed once again, this time for the case when noise in the circuit originates from the equivalent thermal noise source of the input port termination only. It means that all elements of the matrix **C,** except an element corresponding to the input port termination, must equal zero.

According to (4.71) this element is

$$C_{gg} = kT\,(1 - |S_{gg}|^2) \tag{4.132}$$

where

$$S_{gg} = \frac{Z_G - Z_R}{Z_G + Z_R} \tag{4.133}$$

is the reflection coefficient of the input port termination (signal generator port),

Z_G is the impedance of the input port termination (signal generator impedance), and

Z_R is the reference impedance of the signal source port.

$$\mathbf{C}_G = g\begin{array}{c} \\ \end{array}\!\!\overset{\displaystyle g}{\begin{bmatrix} \mathbf{0} & & \vdots & & \\ & \mathbf{0} & \vdots & & \\ & & \ddots\;\vdots & & \\ \cdots & \cdots\;\cdots & \mathbf{C}_{gg} & & \\ & & & \ddots\;\mathbf{0} & \\ & & & & \ddots\;\mathbf{0} \end{bmatrix}} \qquad g = \text{signal generator port number} \tag{4.134}$$

Step 5: Relative to the load impedance, the port diagonal element $N_{G\,l}$ of the matrix $\overline{\mathbf{a}\,\mathbf{a}^{+}}$ is computed using (4.127) as

$$N_{G\,l} = \underline{\mathcal{A}}^{+}\, \mathbf{C}_G\, \underline{\mathcal{A}} \tag{4.135}$$

This time a value of this element multiplied by $(1 - |S_{ll}|^2)$ equals P_{NG}.

Step 6: The noise figure F is found from

$$F = 1 + \frac{N_{\text{int}\,l}}{N_{G\,l}} \tag{4.136}$$

using results obtained in Steps 3 and 5.

The matrix $\mathbf{W}$ in (4.131) is the connection scattering matrix of the analyzed circuit, so a conventional circuit analysis in the frequency domain may be carried out in parallel with the noise analysis and the noise figure computation. This allows a simultaneous optimization of an absolutely general topology microwave circuit with respect to both noise figure and any of the conventional network functions.

The inverse matrices $\mathbf{W}^{-1}$ and $(\mathbf{W}^{-1})^{+}$ of the connection scattering matrix $\mathbf{W}$ may be computed very effectively using the sparse matrix technique described in Chapter 6.

Example 4.2

A commonly used microwave amplifier integrated circuits for high-power and broadband applications is the balanced amplifier configuration. As shown in Figure 4.10, a balanced amplifier consists of two 3-dB and 90-deg directional couplers and two conventional two-port transistor amplifiers.

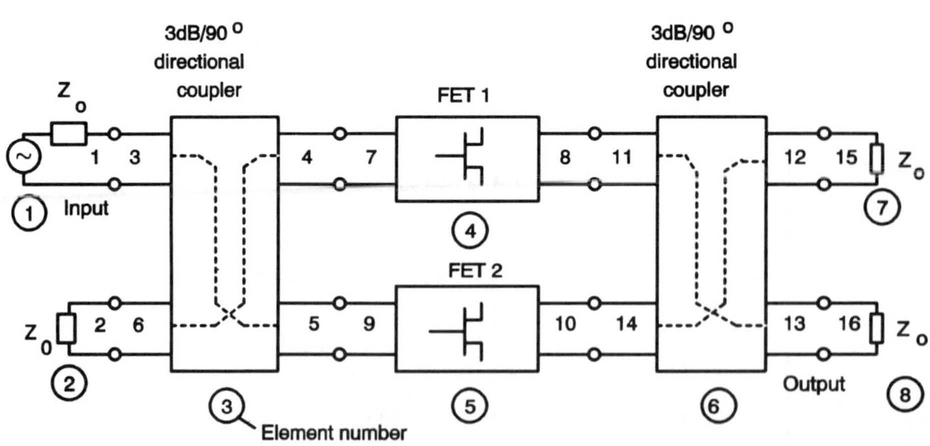

Figure 4.10 Balanced transistor amplifier composed of eight elements (multiports). Elements with numbers 1, 2, 7, and 8 are one-ports; elements with numbers 4 and 5 are two-ports; and elements with numbers 3 and 6 are four-ports. All elements are described by their own scattering matrices and noise wave correlation matrices.

Noise analysis of the sample amplifier is performed in the following way. The connection scattering **W** of the balanced amplifier circuit is

$$
\mathbf{W} = \begin{array}{c} \\ 1 \\ 2 \\ 3 \\ 4 \\ 5 \\ 6 \\ 7 \\ 8 \\ 9 \\ 10 \\ 11 \\ 12 \\ 13 \\ 14 \\ 15 \\ 16 \end{array}
\begin{array}{cccccccccccccccc}
1 & 2 & 3 & 4 & 5 & 6 & 7 & 8 & 9 & 10 & 11 & 12 & 13 & 14 & 15 & 16 \\
-S^{(1)} & 0 & 1 & 0 & 0 & 0 & 0 & 0 & 0 & 0 & 0 & 0 & 0 & 0 & 0 & 0 \\
0 & -S^{(2)} & 0 & 0 & 0 & 1 & 0 & 0 & 0 & 0 & 0 & 0 & 0 & 0 & 0 & 0 \\
1 & 0 & & & & & 0 & 0 & 0 & 0 & 0 & 0 & 0 & 0 & 0 & 0 \\
0 & 0 & & -\mathbf{S}^{(3)} & & & 1 & 0 & 0 & 0 & 0 & 0 & 0 & 0 & 0 & 0 \\
0 & 0 & & & & & 0 & 0 & 1 & 0 & 0 & 0 & 0 & 0 & 0 & 0 \\
0 & 1 & & & & & 0 & 0 & 0 & 0 & 0 & 0 & 0 & 0 & 0 & 0 \\
0 & 0 & 0 & 1 & 0 & 0 & -\mathbf{S}^{(4)} & & 0 & 0 & 0 & 0 & 0 & 0 & 0 & 0 \\
0 & 0 & 0 & 0 & 0 & 0 & & & 0 & 0 & 1 & 0 & 0 & 0 & 0 & 0 \\
0 & 0 & 0 & 0 & 1 & 0 & 0 & 0 & -\mathbf{S}^{(5)} & & 0 & 0 & 0 & 0 & 0 & 0 \\
0 & 0 & 0 & 0 & 0 & 0 & 0 & 0 & & & 0 & 0 & 0 & 1 & 0 & 0 \\
0 & 0 & 0 & 0 & 0 & 0 & 0 & 1 & 0 & 0 & & & & & 0 & 0 \\
0 & 0 & 0 & 0 & 0 & 0 & 0 & 0 & 0 & 0 & & -\mathbf{S}^{(6)} & & & 1 & 0 \\
0 & 0 & 0 & 0 & 0 & 0 & 0 & 0 & 0 & 0 & & & & & 0 & 1 \\
0 & 0 & 0 & 0 & 0 & 0 & 0 & 0 & 0 & 1 & & & & & 0 & 0 \\
0 & 0 & 0 & 0 & 0 & 0 & 0 & 0 & 0 & 0 & 0 & 1 & 0 & 0 & -S^{(7)} & 0 \\
0 & 0 & 0 & 0 & 0 & 0 & 0 & 0 & 0 & 0 & 0 & 0 & 1 & 0 & 0 & -S^{(8)}
\end{array}
$$

In **W**, $\mathbf{S}^{(3)}$ and $\mathbf{S}^{(6)}$ are the scattering matrices of 3-dB/90-deg directional couplers defined in Table 2.1. $\mathbf{S}^{(4)}$ and $\mathbf{S}^{(5)}$ are the scattering matrices of two conventional-type transistor amplifiers. $S^{(1)}$ is the reflection coefficient of the signal generator port. $\mathbf{S}^{(6)}$ is the reflection coefficient of the load impedance Z_L. According to the algorithm for noise computation described in Section 4.5.1, the elements $N_{\text{int }16}$ and $N_{G\ 16}$ (16 is the number of the load port) must be computed by applying relations (4.131) and (4.134). Using matrix $\mathbf{C}_{\text{int}}$ given as

$$
\mathbf{C}_{\mathrm{int}} =
\begin{array}{c}
\\ 1 \\ 2 \\ 3 \\ 4 \\ 5 \\ 6 \\ 7 \\ 8 \\ 9 \\ 10 \\ 11 \\ 12 \\ 13 \\ 14 \\ 15 \\ 16
\end{array}
\begin{array}{cccccccccccccccc}
1 & 2 & 3 & 4 & 5 & 6 & 7 & 8 & 9 & 10 & 11 & 12 & 13 & 14 & 15 & 16 \\
0 & 0 & 0 & 0 & 0 & 0 & 0 & 0 & 0 & 0 & 0 & 0 & 0 & 0 & 0 & 0 \\
0 & C_s^{(2)} & 0 & 0 & 0 & 0 & 0 & 0 & 0 & 0 & 0 & 0 & 0 & 0 & 0 & 0 \\
0 & 0 & & & & & 0 & 0 & 0 & 0 & 0 & 0 & 0 & 0 & 0 & 0 \\
0 & 0 & \multicolumn{4}{c}{\mathbf{C}_s^{(3)}} & 0 & 0 & 0 & 0 & 0 & 0 & 0 & 0 & 0 & 0 \\
0 & 0 & & & & & 0 & 0 & 0 & 0 & 0 & 0 & 0 & 0 & 0 & 0 \\
0 & 0 & & & & & 0 & 0 & 0 & 0 & 0 & 0 & 0 & 0 & 0 & 0 \\
0 & 0 & 0 & 0 & 0 & 0 & \multicolumn{2}{c}{\mathbf{C}_s^{(4)}} & 0 & 0 & 0 & 0 & 0 & 0 & 0 & 0 \\
0 & 0 & 0 & 0 & 0 & 0 & & & 0 & 0 & 0 & 0 & 0 & 0 & 0 & 0 \\
0 & 0 & 0 & 0 & 0 & 0 & 0 & 0 & \multicolumn{2}{c}{\mathbf{C}_s^{(5)}} & 0 & 0 & 0 & 0 & 0 & 0 \\
0 & 0 & 0 & 0 & 0 & 0 & 0 & 0 & & & 0 & 0 & 0 & 0 & 0 & 0 \\
0 & 0 & 0 & 0 & 0 & 0 & 0 & 0 & 0 & 0 & & & & & 0 & 0 \\
0 & 0 & 0 & 0 & 0 & 0 & 0 & 0 & 0 & 0 & & & & & 0 & 0 \\
0 & 0 & 0 & 0 & 0 & 0 & 0 & 0 & 0 & 0 & \multicolumn{4}{c}{\mathbf{C}_s^{(6)}} & 0 & 0 \\
0 & 0 & 0 & 0 & 0 & 0 & 0 & 0 & 0 & 0 & & & & & 0 & 0 \\
0 & 0 & 0 & 0 & 0 & 0 & 0 & 0 & 0 & 0 & 0 & 0 & 0 & 0 & C_s^{(7)} & 0 \\
0 & 0 & 0 & 0 & 0 & 0 & 0 & 0 & 0 & 0 & 0 & 0 & 0 & 0 & 0 & 0
\end{array}
$$

and relations (4.131), we may compute the noise power at the output port of the amplifier that originated from the noise sources acting within the circuit. In $\mathbf{C}_{\mathrm{int}}$, $\mathbf{C}_s^{(3)}$ and $\mathbf{C}_s^{(6)}$ are noise wave correlation matrices of two directional couplers, whereas $\mathbf{C}_s^{(4)}$ and $\mathbf{C}_s^{(5)}$ describe noise properties of two transistor amplifiers. It should be noted here that the elements $C_{1,1}$ and $C_{16,16}$ of the $\mathbf{C}_{\mathrm{int}}$ matrix are set to zero (1 is the number of signal generator port and 16 is the number of the load impedance port of the amplifier). The noise wave correlation matrices of passive elements occurring in the analyzed circuit are computed using (4.71).

The matrix $\mathbf{C}_G$ given below as

$$
\mathbf{C}_G = \begin{array}{c}
\\ 1 \\ 2 \\ 3 \\ 4 \\ 5 \\ 6 \\ 7 \\ 8 \\ 9 \\ 10 \\ 11 \\ 12 \\ 13 \\ 14 \\ 15 \\ 16
\end{array}
\begin{array}{cccccccccccccccc}
1 & 2 & 3 & 4 & 5 & 6 & 7 & 8 & 9 & 10 & 11 & 12 & 13 & 14 & 15 & 16 \\
C_s^{(1)} & 0 & 0 & 0 & 0 & 0 & 0 & 0 & 0 & 0 & 0 & 0 & 0 & 0 & 0 & 0 \\
0 & 0 & 0 & 0 & 0 & 0 & 0 & 0 & 0 & 0 & 0 & 0 & 0 & 0 & 0 & 0 \\
0 & 0 & 0 & 0 & 0 & 0 & 0 & 0 & 0 & 0 & 0 & 0 & 0 & 0 & 0 & 0 \\
0 & 0 & 0 & 0 & 0 & 0 & 0 & 0 & 0 & 0 & 0 & 0 & 0 & 0 & 0 & 0 \\
0 & 0 & 0 & 0 & 0 & 0 & 0 & 0 & 0 & 0 & 0 & 0 & 0 & 0 & 0 & 0 \\
0 & 0 & 0 & 0 & 0 & 0 & 0 & 0 & 0 & 0 & 0 & 0 & 0 & 0 & 0 & 0 \\
0 & 0 & 0 & 0 & 0 & 0 & 0 & 0 & 0 & 0 & 0 & 0 & 0 & 0 & 0 & 0 \\
0 & 0 & 0 & 0 & 0 & 0 & 0 & 0 & 0 & 0 & 0 & 0 & 0 & 0 & 0 & 0 \\
0 & 0 & 0 & 0 & 0 & 0 & 0 & 0 & 0 & 0 & 0 & 0 & 0 & 0 & 0 & 0 \\
0 & 0 & 0 & 0 & 0 & 0 & 0 & 0 & 0 & 0 & 0 & 0 & 0 & 0 & 0 & 0 \\
0 & 0 & 0 & 0 & 0 & 0 & 0 & 0 & 0 & 0 & 0 & 0 & 0 & 0 & 0 & 0 \\
0 & 0 & 0 & 0 & 0 & 0 & 0 & 0 & 0 & 0 & 0 & 0 & 0 & 0 & 0 & 0 \\
0 & 0 & 0 & 0 & 0 & 0 & 0 & 0 & 0 & 0 & 0 & 0 & 0 & 0 & 0 & 0 \\
0 & 0 & 0 & 0 & 0 & 0 & 0 & 0 & 0 & 0 & 0 & 0 & 0 & 0 & 0 & 0 \\
0 & 0 & 0 & 0 & 0 & 0 & 0 & 0 & 0 & 0 & 0 & 0 & 0 & 0 & 0 & 0 \\
0 & 0 & 0 & 0 & 0 & 0 & 0 & 0 & 0 & 0 & 0 & 0 & 0 & 0 & 0 & 0
\end{array}
$$

is used to compute the noise power at the output port of the amplifier originating from the thermal noise of the signal generator impedance Z_G. This time, in the matrix $\mathbf{C}_G$ only element $C_{11} = kT\,(1 - |S_{11}|^2)$ is the nonzero element (signal source port is numbered as one).

Sparse matrix techniques described in the next chapter are used to compute (4.135).

4.7.2 The Algorithm for Computation of Four Noise Parameters of a Two-Port With any Internal Topology

To compute four noise parameters related to the input and output ports of a general multiport circuit shown in Figure 4.9 it is assumed that the reflection coefficient S_{gg} of the input port termination (4.133) and the reflection coefficient S_{ll} of the output port load (4.124) are both zero. It is assumed also that both these terminations are noisefree. Under such an assumption, the noise waves at the input and output ports of the overall circuit can be computed using (4.120) [24].

The actual steps carried out by the numerical algorithm are as follows.

Step 1: The connection scattering matrix **W** of the analyzed circuit is built and computed. Diagonal elements $W_{gg} = -S_{gg}$ and $W_{ll} = -S_{ll}$ of **W** must be set to zero (where g denotes signal generator port number and l the load impedance port number).

Step 2: The noise correlation matrices of the individual passive and active circuit elements are found. The noise correlation matrix **C** defined by (4.121) is built and computed. Two diagonal elements of matrix **C** relative to ports belonging to the signal source impedance and to the load impedance are set to zero.

Step 3: Three elements

$$\overline{(\mathbf{a}\,\mathbf{a}^+)}_{gg},\ \overline{(\mathbf{a}\,\mathbf{a}^+)}_{ll},\ \text{and}\ \overline{(\mathbf{a}\,\mathbf{a}^+)}_{gl}\quad (\text{or}\ \overline{(\mathbf{a}\,\mathbf{a}^+)}_{lg})$$

of the matrix $\overline{\mathbf{a}\,\mathbf{a}^+}$ are computed using the following relations.

For the element $\overline{(\mathbf{a}\,\mathbf{a}^+)}_{gg}$,

$$\overline{(\mathbf{a}\,\mathbf{a}^+)}_{gg} = \mathbf{e}_g^T\,\overline{\mathbf{a}\,\mathbf{a}^+}\,\mathbf{e}_g \tag{4.137}$$

where $\mathbf{e}_g$ is a vector whose elements are all zeros except 1 in the position g corresponding to the generator port. Using (4.127), equation (4.137) may be rewritten in the form

$$\overline{(\mathbf{a}\,\mathbf{a}^+)}_{gg} = \mathbf{e}_g^T\,\mathbf{W}^{-1}\,\mathbf{C}\,(\mathbf{W}^+)^{-1}\,\mathbf{e}_g = [(\mathbf{W}^+)^{-1}\,\mathbf{e}_g]\,\mathbf{C}\,(\mathbf{W}^+)^{-1}\,\mathbf{e}_g = \underline{\mathcal{A}}_g^+\,\mathbf{C}\,\underline{\mathcal{A}}_g \tag{4.138}$$

where the vector

$$\underline{\mathcal{A}}_g = (\mathbf{W}^+)^{-1}\,\mathbf{e}_g = (\mathbf{W}^{T*})^{-1}\,\mathbf{e}_g \tag{4.139}$$

is the solution vector of the system of equations

$$\mathbf{W}^{T*}\,\underline{\mathcal{A}}_g = \mathbf{e}_g \tag{4.140}$$

whose coefficient matrix is equal to the complex transpose of the connection scattering matrix **W** and has $\mathbf{e}_g$ as its right-hand-side vector.

For the element $\overline{(\mathbf{a}\,\mathbf{a}^+)}_{ll}$,

$$\overline{(\mathbf{a}\,\mathbf{a}^+)}_{ll} = \mathbf{e}_l^T\, \overline{\mathbf{a}\,\mathbf{a}^+}\, \mathbf{e}_l \tag{4.141}$$

where $\mathbf{e}_l$ is a vector whose elements are all zeros except 1 in the position l corresponding to the load port number. As in the case of the element $\overline{(\mathbf{a}\,\mathbf{a}^+)}_{gg}$ we have

$$\overline{(\mathbf{a}\,\mathbf{a}^+)}_{ll} = = \underline{\mathcal{A}}_l^+\, \mathbf{C}\, \underline{\mathcal{A}}_l \tag{4.142}$$

where the vector

$$\underline{\mathcal{A}}_l = (\mathbf{W}^+)^{-1}\, \mathbf{e}_l = (\mathbf{W}^{T*})^{-1}\, \mathbf{e}_l \tag{4.143}$$

is the solution vector of the system of equations

$$\mathbf{W}^{T*}\, \underline{\mathcal{A}}_l = \mathbf{e}_l \tag{4.144}$$

whose coefficient matrix is equal to the complex transpose of the connection scattering matrix **W** and has $\mathbf{e}_l$ as its right-hand-side vector.

For the element $\overline{(\mathbf{a}\,\mathbf{a}^+)}_{gl}$,

$$\overline{(\mathbf{a}\,\mathbf{a}^+)}_{gl} = \mathbf{e}_g^T\, \overline{\mathbf{a}\,\mathbf{a}^+}\, \mathbf{e}_l \tag{4.145}$$

Using the relations (4.140), (4.144), and (4.145) we have

$$\overline{(\mathbf{a}\,\mathbf{a}^+)}_{gl} = = \underline{\mathcal{A}}_g^+\, \mathbf{C}\, \underline{\mathcal{A}}_l \tag{4.146}$$

The correlation matrix of the outgoing noise waves c_1 and c_2 at the input and output ports of the overall network is given by (see Figure 4.11)

$$\mathbf{C}_{\text{net}} = \begin{bmatrix} C11 & C12 \\ C12* & C22 \end{bmatrix} = \begin{bmatrix} \overline{c_1\, c_1^*} & \overline{c_1\, c_2^*} \\ \overline{c_2\, c_1^*} & \overline{c_2\, c_2^*} \end{bmatrix}$$

$$= \begin{bmatrix} \overline{(\mathbf{a}\,\mathbf{a}^+)}_{gg} & \overline{(\mathbf{a}\,\mathbf{a}^+)}_{gl} \\ \overline{(\mathbf{a}\,\mathbf{a}^+)}_{gl}^* & \overline{(\mathbf{a}\,\mathbf{a}^+)}_{ll} \end{bmatrix} \tag{4.147}$$

Once the elements of the noise correlation matrix $\mathbf{C}_{\text{net}}$ are known, any set of

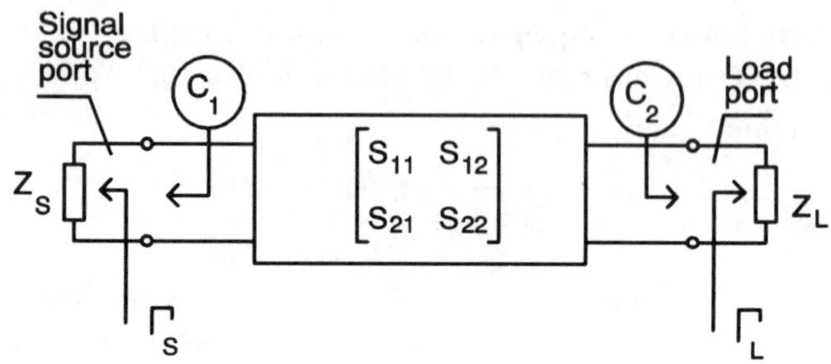

Figure 4.11 Resultant noise waves c_1 and c_2 at the input and output ports of the overall network. Correlation matrix of these noise waves describes noise properties of the entire network.

noise parameters of the overall two-port can be computed. We can compute, for example, F_{min}, Γ_{opt}, and N parameters using relations (4.100), (4.101), (4.102), and (4.103). The noise figure of the analyzed circuit can be computed now in two ways. First, the noise figure F of the overall circuit can be computed using four noise parameters F_{min}, Γ_{opt}, and N and (4.95)

$$F = F_{min} + N\frac{|\Gamma_G - \Gamma_{opt}|^2}{(1 - |\Gamma_G|^2)(1 - |\Gamma_{opt}|^2)} \tag{4.148}$$

Because the four noise parameters F_{min}, Γ_{opt}, and N are equivalent to the elements $C11$, $C12$, and $C22$ of the overall noise correlation matrix, the second way to compute the noise figure F of the circuit is by using the relation

$$F = 1 + \frac{C11\left|\dfrac{S21\,\Gamma_G}{1 - S11\,\Gamma_G}\right|^2 + C22 + 2\,\mathrm{Re}\left\{C12\dfrac{S21\,\Gamma_G}{1 - S11\,\Gamma_G}\right\}}{kT_0\,(1 - |\Gamma_G|^2)\left|\dfrac{S21}{1 - S11\,\Gamma_G}\right|^2} \tag{4.149}$$

where $S11$ and $S21$ are the scattering parameters of the overall circuit and

$$\Gamma_G = \frac{Z_G - Z_R}{Z_G + Z_R} \tag{4.150}$$

is the reflection coefficient of the signal source impedance with respect to the input port reference impedance Z_R.

The resulting overall two-port scattering parameters $S11$ and $S21$ used in (4.149) are computed using the method discussed in Section 3.2.3.

4.8 CONSTANT NOISE FIGURE CIRCLES

From Sections 4.6 and 4.7 it is known that noise figure F is a function of the generator reflection coefficient Γ_G presented to the input port of the noisy two-port by the signal generator. The noise figure does not depend on Γ_L. It may be proved that there is a range of values of the generator port reflection coefficient over which the noise figure is constant. For a given constant value F_i of the noise figure F, the solution for Γ_G lies on a circle [26].

By introducing the parameter

$$N_i = \frac{F_i - F_{\text{min}}}{4\ R_n/Z_R} |1 + \Gamma_{G\,\text{opt}}|^2 \tag{4.151}$$

it may be shown that the center and the radius of the constant noise figure circle are expressed by

$$\text{the center} \quad c_{F_i} = \frac{\Gamma_{G\,\text{opt}}}{1 + N_i} \tag{4.152}$$

$$\text{the radius} \quad r_{F_i} = \frac{1}{1 + N_i} \sqrt{N_i^2 + N_i\,(1 - |\Gamma_{G\,\text{opt}}|^2)} \tag{4.153}$$

It can be seen from these equations that $N_i = 0$ corresponds to the F_{min} circle with zero radius. This point F_{min} is located at $\Gamma_{G\ \text{opt}}$ on the Smith chart. The centers of other noise figure circles lie along the $\Gamma_{G\ \text{opt}}$ vector.

4.9 CONSTANT NOISE MEASURE CIRCLES

The noise measure of the two-port network may be described in terms of the signal generator reflection coefficient [26, 27]

$$M = \frac{F_e\,(1 - |\Gamma_G|^2)\,|S_{21}|^2|1 + \Gamma_{\text{opt}}|^2 + 4\,\dfrac{R_n}{Z_R}\,|S_{21}|^2|\Gamma_G - \Gamma_{\text{opt}}|^2}{|1 + \Gamma_{\text{opt}}|^2\,[|S_{21}|^2\,(1 - |\Gamma_G|^2) - (1 - |S_{22}|^2) - |\Gamma_G|^2\,(|s_{11}|^2 - |\Delta|^2) + 2\,\text{Re}(\Gamma_G C_1)]} \tag{4.154}$$

It can be shown that (4.154) represents circles in the signal source reflection coefficient plane Γ_G. The centers and the radius of the constant noise measure circles are

$$C_m = \frac{M\,|1+\Gamma_{opt}|^2\, C_1^* + 4\,\dfrac{R_n}{Z_R}\,|S_{21}|^2\,\Gamma_{opt}}{M\,|1+\Gamma_{opt}|^2\, P + |S_{21}|^2 \left(4\,\dfrac{R_n}{Z_r} - W\right)} \tag{4.155}$$

$$r_m = \frac{\sqrt{M^2\,M_1 + M\,M_2 + M_3}}{M\,|1+\Gamma_{opt}|^2\, P + |S_{21}|^2 \left(4\,\dfrac{R_n}{Z_R} - W\right)} \tag{4.156}$$

where

$$P = |S_{21}|^2 + |S_{11}|^2 - |\Delta|^2 \tag{4.157}$$

$$Q = |S_{21}|^2 + |S_{11}|^2 - 1 \tag{4.158}$$

$$W = |1+\Gamma_{opt}|^2\, F_{emin} \tag{4.159}$$

$$M_1 = |1+\Gamma_{opt}|^2\,(P\,Q + |C_1|^2) \tag{4.160}$$

$$M_2 = |1+\Gamma_{opt}|^2\,|S_{21}|^2 \left[8\,\frac{R_n}{Z_R}\,\mathrm{Re}\{\Gamma_{opt}\,C_1\} - \left(4\,\frac{R_n}{Z_R}\,|\Gamma_{opt}|^2 + W\right)P - \left(W - 4\,\frac{R_n}{Z_R}\right)Q\right] \tag{4.161}$$

$$M_3 = |S_{21}|^2 \left[W - 4\,\frac{R_n}{Z_R}\,(1 - |\Gamma_{opt}|^2)\right] \tag{4.162}$$

$$C_1 = S_{11} - S_{22}^*\,(S_{11}S_{22} - S_{12}S_{21}) \tag{4.163}$$

The minimum value of the noise measure corresponds to the noise measure circle of zero radius. From (4.156), by setting $r_m = 0$, we have

$$M_{min} = \frac{-M_2 \pm \sqrt{M_1^2 - 4\,M_1\,M_3}}{2\,M_1} \tag{4.164}$$

The minimum value of M corresponds to the optimum value of the source reflection coefficient, which can be obtained from (4.155), and is given by

$$\Gamma_{\text{M opt}} = \frac{M_{\min} |1 + \Gamma_{\text{opt}}|^2 C_1^* + 4 \frac{R_n}{Z_R} |S_{21}| \Gamma_{\text{opt}}}{M_{\min} |1 + \Gamma_{\text{opt}}|^2 P + |S_{21}|^2 \left(4 \frac{R_n}{Z_R} - W\right)} \tag{4.165}$$

Minimum noise measure $M_{\min}$ is an invariant parameter and is not affected by a lossless embedding network.

References

[1] Dobrowolski, J. A., *Introduction to Computer Methods for Microwave Circuit Analysis and Design,* Norwood, MA: Artech House, 1991.

[2] Rothe, H., and W. Dahlke, "Theory of Noisy Fourpoles," *Proc. IRE,* Vol. 44, June 1956, pp. 811–818.

[3] Bauer, H., and H. Rothe, "Der equivalente Rauschvierpol als Wellenvierpol," *Archiv der Elektrischen Uebertragung,* Vol. 10, 1956, pp. 241–252.

[4] Kurokawa, K., "Power Waves and the Scattering Matrix," *IEEE Trans. Microwave Theory Tech.,* Vol. MTT-13, March 1965, pp. 194–202.

[5] Hillbrandt, H., and P. H. Russer, "An Efficient Method for Computer Aided Noise Analysis of Linear Amplifier Networks," *IEEE Trans. Circuits Syst.,* Vol. CAS-23, April 1976, pp. 235–238.

[6] Reiss, R., "Four-Pole Based Noise Analysis by Digital Computer," *Proc. 1976 European Conf. on Circuit Theory and Design,* Genova, Italy, Sept. 7–10, 1976, pp. 632–642.

[7] Hause, H. A., and R. B. Adler, *Circuit Theory of Linear Noisy Networks,* New York: John Wiley, 1959.

[8] Bosma, H., "On the Theory of Linear Noisy Systems," *Philips Res. Repts. Suppl.,* No. 10, 1967.

[9] Meys, R. P., "A Wave Approach to the Noise Properties of Linear Microwave Devices," *IEEE Trans. Microwave Theory Tech.,* Vol. MTT-26, January 1978, pp. 34–37.

[10] Grosch, T. O., and L. A. Carpenter, "Two-port to Three-Port Noise Wave Transformation for CAD Applications," *IEEE Trans. Microwave Theory Tech.,* Vol. MTT-41, September 1993, pp. 1543–1548.

[11] Wedge, S. W., and D. B. Rutledge, "Wave Techniques for Noise Modeling and Measurement," *IEEE Trans. Microwave Theory Tech.,* Vol. 40, No. 11, November 1992, pp. 2004–2012.

[12] Wedge, S. W., and D. B. Rutledge, "Noise Waves and Passive Linear Multiports," *IEEE Microwave Guided Wave Lett.,* Vol. 1., May 1991, pp. 117–119.

[13] Statz, H., H. A. Haus, and R. A. Pucel, "Noise Characteristics of Gallium Arsenide Field-Effect Transistors," *IEEE Trans. Electron Devices,* Vol. ED-21, September 1974, pp. 549–562.

[14] Pucel, R. A., D. J. Masse, and C. K. Krumm, "Noise Performance of Gallium Arsenide Field-Effect Transistors," *IEEE J. Solid-State Circuits,* Vol. SC-11, April 1976, pp. 243–255.

[15] Pudell, A. F., "A Functional GaAs FET Noise Model," *IEEE Trans. Electron Devices,* Vol. ED-28, May 1981, pp. 511–517.

[16] Gupta, M. S., O. Pitzalis, S. E. Rosenbaum, and P. T. Greiling, "Microwave Noise Characterization of GaAs MESFET's: Evaluation by on Wafer Low Frequency Output Noise Current Measurement," *IEEE Trans. Microwave Theory Tech.,* Vol. MTT-35, December 1987, pp. 1208–1218.

[17] Cappy, A., "Noise Modeling and Measurement Techniques," *IEEE Trans. Microwave Theory Tech.,* Vol. 36, January 1988, pp. 1–10.

[18] Froelich, R. K., "An Improved Model for Noise Characterization of Microwave GaAs FET's," *IEEE Trans. Microwave Theory Tech.,* Vol. MTT-38, June 1990, pp. 703–706.

[19] Pospieszalski, M. W., Modeling of Noise Parameters of MESFET's and MODFET's and their Fre-

quency and Temperature Dependence," *IEEE Trans. Microwave Theory Tech.,* Vol. 37, September 1989, pp. 1340–1350.

[20] Lange, J., "Noise Characterization of Linear Two-Ports in Terms of Invariant Parameters," *IEEE J. Solid-State Circuits,* Vol. SC-2, June 1967, pp. 37–40.

[21] Wiatr, W., *A Method of Estimating Noise Parameters of Linear Microwave Two-Ports,* Ph.D. dissertation, Warsaw Univ. of Technology, Warsaw, Poland, 1980. (Polish)

[22] Lane, R. Q., "The Determination of the Device Noise Parameters," *Proc. IEEE,* Vol. 116, August 1969, pp. 1461–1462.

[23] Caruso, G., and M. Sannino, "Computer-Aided Determination of Microwave Two-Port Noise Parameters," *IEEE Trans. Microwave Theory Tech.,* Vol. MTT-26, September 1978, pp. 639–642.

[24] Dobrowolski, J. A., "A CAD Oriented Method for Noise Figure Computation of Two-Ports With Any Internal Topology," *IEEE Trans. Microwave Theory Tech.*, Vol. MTT-37, January 1989, pp. 15–20.

[25] Hecken, R. P., "Analysis of Linear Noisy Two-Ports Using Scattering Waves," *IEEE Trans. Microwave Theory Tech.*, Vol. MTT-29, October 1981, pp. 997–1003.

[26] Fukui, H., "Available Power Gain, Noise Figure, and Noise Measure of Two-Ports and Their Graphical Representation," *IEEE Trans. Circuit Theory,* June 1966, pp. 137–142.

[27] Vendelin, G. D., A. M. Pavio, and U. L. Rohde, *Microwave Circuit Design Using Linear and Nonlinear Techniques,* New York: J. Wiley & Sons, 1990.

CHAPTER 5

Microwave Network Sensitivity Analysis Using the Wave Approach

Information about sensitivities or partial derivatives of certain network functions with respect to network parameters is very important for the design of microwave networks. There are many applications of sensitivities in the computer-aided design of microwave networks. First, sensitivities allow the designer to understand how network parameter changes affect network performance. Knowledge of sensitivities is the key to identifying the network parameters that have the most influence on network performance. Second, sensitivity analysis in connection with the optimization methods allows a more algorithmic design procedure and, consequentially, automation of the circuit design. In this application, sensitivities are used to determine gradients of network functions. Third, sensitivities can be used to choose the best network among several networks that all have the same specific performance. In general, we assume that a network with the least performance sensitivities with respect to network parameters is the best one. Sensitivities are also used in tolerance analysis and design procedures and in yield-driven procedures for the design of microwave networks. All these problems are discussed in detail elsewhere [1–4].

In Section 5.1, sensitivity analysis based on the transposed scattering matrix method is discussed. The results of network sensitivity analysis are sensitivities of the overall circuit scattering parameters with respect to specified network parameters.

In Section 5.2, noise power sensitivities of microwave networks are presented. They can be applied to efficient noise figure gradient computation. The approach is applicable to networks with any internal topology.

Scattering parameters are complex functions, and their sensitivities are considered in the following way. Since any complex quantity may be expressed in the form

$$f = |f| \, e^{j\phi} \tag{5.1}$$

taking the natural logarithm as

$$\ln f = \ln|f| + j\phi \tag{5.2}$$

and differentiating with respect to a parameter p we obtain

$$\frac{1}{f}\frac{\partial f}{\partial p} = \frac{1}{|f|}\frac{\partial |f|}{\partial p} + j\frac{\partial \phi}{\partial p} \tag{5.3}$$

This complex equation can be split into two equations as

$$\frac{\partial |f|}{\partial p} = |f| \operatorname{Re}\left\{\frac{1}{f}\frac{\partial f}{\partial p}\right\} \tag{5.4}$$

and

$$\frac{\partial \phi}{\partial p} = \operatorname{Im}\left\{\frac{1}{f}\frac{\partial f}{\partial p}\right\} \tag{5.5}$$

where Re and Im denote the real and imaginary parts.

Very often the magnitude (the absolute value) is measured in decibels (dB). To obtain the corresponding formula we write (5.2) in the form

$$\ln f = A + j\varphi \tag{5.6}$$

where $A = \ln|f|$.

Differentiation with respect to p gives

$$\frac{1}{f}\frac{\partial f}{\partial p} = \frac{\partial A}{\partial p} + j\frac{\partial \phi}{\partial p} \tag{5.7}$$

The sensitivity $\partial A/\partial p$ is expressed in nepers. In order to convert it to decibels we multiply the value of $\partial A/\partial p$ by $20 \cdot \log_{10} e = 8.686$, which is

$$\frac{\partial A}{\partial p}\,[\mathrm{dB}] = 8.686\,\frac{\partial A}{\partial p} = 8.686 \operatorname{Re}\left\{\frac{1}{f}\frac{\partial f}{\partial p}\right\} \tag{5.8}$$

5.1 TRANSPOSED MATRIX METHOD FOR SENSITIVITY ANALYSIS

The computational method for determining sensitivities of microwave networks described by the connection scattering matrix is based on the following considerations [5–8].

Differentiating

$$\mathbf{W}\,\mathbf{a} = \mathbf{c} \tag{5.9}$$

with respect to the parameter p we have

$$\frac{\partial \mathbf{a}}{\partial p} = -\mathbf{W}^{-1}\frac{\partial \mathbf{W}}{\partial p}\mathbf{W}^{-1}\,\mathbf{c} = \mathbf{W}^{-1}\frac{\partial \mathbf{S}}{\partial p}\,\mathbf{a} \tag{5.10}$$

In (5.10) we assume $\partial \mathbf{c}/\partial p = 0$.

The sensitivity of the ingoing wave variable a_r at the rth port of the network may be computed by multiplying the left-hand side of (5.10) by a row vector $\mathbf{e}_r^{\mathrm{T}}$ all of whose elements are equal to null except for the rth element that is equal to 1; that is,

$$\mathbf{e}_r^{\mathrm{T}} = [0, 0, \ldots, 0, 1, 0, \ldots, 0] \tag{5.11}$$

Performing this operation yields

$$\frac{\partial a_r}{\partial p} = \mathbf{e}_r^{\mathrm{T}}\frac{\partial \mathbf{a}}{\partial p} = \mathbf{e}_r^{\mathrm{T}}\,\mathbf{W}^{-1}\frac{\partial \mathbf{S}}{\partial p}\,\mathbf{a} = [(\mathbf{W}^{\mathrm{T}})^{-1}\mathbf{e}_r]^{\mathrm{T}}\frac{\partial \mathbf{S}}{\partial p}\,\mathbf{a} = \alpha^{\mathrm{T}}\frac{\partial \mathbf{S}}{\partial p}\,\mathbf{a} \tag{5.12}$$

where

$$\alpha = (\mathbf{W}^{\mathrm{T}})^{-1}\,\mathbf{e}_r = \hat{\mathbf{W}}^{-1}\,\mathbf{e}_r \tag{5.13}$$

is a solution vector of the transposed or adjoint matrix equation

$$\mathbf{W}^{\mathrm{T}}\,\alpha = \mathbf{e}_r \tag{5.14}$$

The above matrix equation may be treated as the equation of the adjoint network excited at the rth port by a signal generator with the impressed wave equal to 1. According to (5.14) the connection scattering matrix $\hat{\mathbf{W}}$ of the adjoint network must be equal to the transposed connection scattering matrix $\mathbf{W}$ of the original network.

Because

$$\hat{\mathbf{W}} = \mathbf{W}^{\mathrm{T}} = (\mathbf{\Gamma} - \mathbf{S})^{\mathrm{T}} = \mathbf{\Gamma} - \mathbf{S}^{\mathrm{T}} \tag{5.15}$$

it means that topologies of the original and the adjoint networks are the same, whereas the elements of the adjoint network have the scattering matrices $\hat{\mathbf{s}}^{(1)}$, $\hat{\mathbf{s}}^{(2)}$, . . . , $\hat{\mathbf{s}}^{(m)}$ equal to the transposed scattering matrices of the corresponding elements of the original network.

The general relation (5.12), which defines the sensitivities of the ingoing wave variable at the rth port a_r with respect to many network parameters, has the form

$$\mathbf{G} = \nabla a_r = \begin{bmatrix} \alpha^{\mathrm{T}} \dfrac{\partial \mathbf{S}}{\partial p_1} \mathbf{a} \\ \alpha^{\mathrm{T}} \dfrac{\partial \mathbf{S}}{\partial p_2} \mathbf{a} \\ \vdots \\ \alpha^{\mathrm{T}} \dfrac{\partial \mathbf{S}}{\partial p_n} \mathbf{a} \end{bmatrix} = \begin{bmatrix} \sum_{k \in E_1} \alpha^{(k)\mathrm{T}} \dfrac{\partial \mathbf{S}^{(k)}}{\partial p_1} \mathbf{a}^{(k)} \\ \sum_{k \in E_2} \alpha^{(k)\mathrm{T}} \dfrac{\partial \mathbf{S}^{(k)}}{\partial p_2} \mathbf{a}^{(k)} \\ \vdots \\ \sum_{k \in E_n} \alpha^{(k)\mathrm{T}} \dfrac{\partial \mathbf{S}^{(k)}}{\partial p_n} \mathbf{a}^{(k)} \end{bmatrix} \tag{5.16}$$

where $E_1, E_2, \ldots, E_n$ represent sets of circuit elements that depend on the parameters $p_1, p_2, \ldots p_n$, respectively.

For each parameter p_i the matrix $\partial \mathbf{S}/\partial p_i$ is formed and the products indicated on the right-hand side of (5.16) must be evaluated. As the vectors **a** and α are independent of the parameter index i, we see that the application of (5.16) requires the solution of only two sets of linear equations, (5.11) and (5.14), irrespective of the number of parameters p_i.

5.2 SENSITIVITY COMPUTATION OF THE OVERALL SCATTERING PARAMETERS

The proposed method may be used effectively to compute **S** parameter sensitivities with respect to any independent variable parameter p.

Let us assume that we analyze a network for which we wish to compute the transfer function $S_{ij} = |S_{ij}|\exp(j\phi_{ij})$ and then evaluate its sensitivities. If the jth port of the network is excited by a matched generator with impressed wave $c = 1$ and all external ports are terminated in matched loads, then the transfer function S_{ij} is given by (for $i \neq j$)

$$S_{ij} = b_i \big|_{a_j=1} = a_l \tag{5.17}$$

In (5.17) a_l is the ingoing wave variable at the lth port of the matched load terminating the ith port of the network. The condition $a_j = 1$ is imposed by the signal generator with $c = 1$ connected to port j.

Now, if the adjoint network with $\hat{\mathbf{W}} = \mathbf{W}^{\mathrm{T}}$ is excited at port l by a matched generator with impressed wave $e_l = 1$ ($\alpha_i = 1$), and because $\partial S_{ij}/\partial p = \partial a_l/\partial p$, we obtain

$$\frac{\partial S_{ij}}{\partial p} = \alpha^{\mathrm{T}} \frac{\partial \mathbf{S}}{\partial p} \mathbf{a} = \sum_{k \in E} \alpha^{(k)\mathrm{T}} \frac{\partial \mathbf{S}^{(k)}}{\partial p} \mathbf{a}^{(k)} \tag{5.18}$$

The sensitivities of any S_{ij} are determined by suitably exciting the original network ($\alpha_j = 1$) and the adjoint network ($\alpha_i = 1$). In (5.18) the summation is taken over all network elements that are dependent upon the parameter p.

Figure 5.1 illustrates network models for the sensitivity computation of transfer function S_{ij} of a microwave network.

The original network in Figure 5.1(a) corresponds to the equation

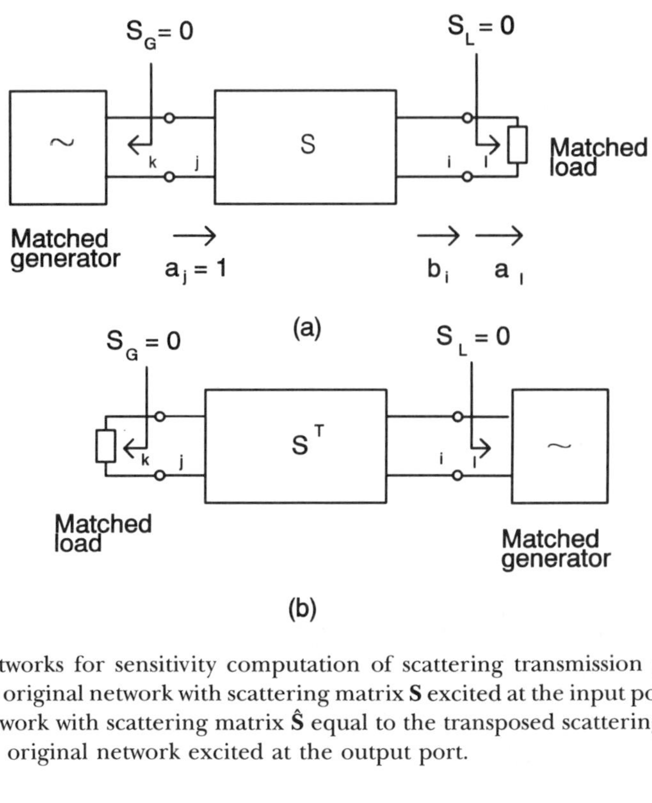

Figure 5.1 Networks for sensitivity computation of scattering transmission parameter S_{ji}: (a) original network with scattering matrix $\mathbf{S}$ excited at the input port; (b) adjoint network with scattering matrix $\hat{\mathbf{S}}$ equal to the transposed scattering matrix $\mathbf{S}^{\mathrm{T}}$ of the original network excited at the output port.

$$\mathbf{W}\,\mathbf{a} = \mathbf{c} \tag{5.19}$$

in which elements $w_{kk} = S_G$ and $w_{ll} = S_L$ are set to zero. This is because the scattering parameter S_G of the generator with the port number k, connected to the input port number j, must be zero (matched generator). Also the scattering parameter S_L of the load with the port numbered l, connected to the output port i, must be zero (matched load). Vector $\mathbf{c}$ must have all elements equal to 0, except the kth element set to 1.

The adjoint network in Figure 5.1(b) corresponds to the equation

$$\mathbf{W}^T\,\mathbf{a} = \mathbf{e}_l \tag{5.20}$$

In vector $\mathbf{e}_l$ except for the lth element that is set to 1, all elements are equal to 0.

Table 5.1 presents scattering matrix partial derivatives with respect to element parameters for basic network elements.

5.3 GROUP DELAY COMPUTATION OF MICROWAVE NETWORK TRANSMISSION FUNCTIONS

A characteristic of great importance for some microwave networks such as linear amplifiers and filters is the group delay versus frequency of transmission functions.

In the case of microwave networks described by the scattering matrices the group delay is defined as

$$\tau_{ij} = -\frac{\partial \phi_{ij}}{\partial \omega} \tag{5.21}$$

where $\phi_{ij} = \arg\{\,S_{ij}\,\}$ and S_{ij} is the transmission scattering parameter between the output port i and the input port j.

We may compute τ_{ij} by means of the incremental ratio

$$\tau_{ij} = -\frac{\Delta \phi_{ij}}{\Delta \omega} \tag{5.22}$$

If the increment $\Delta\omega$ is not small enough, a great computational error can be made while computing τ_{ij}. The incremental values $\Delta\phi_{ij}/\Delta\omega$ tend to the true value of τ_{ij} only in the limit as $\Delta\omega \to 0$, but a very small value of $\Delta\omega$ is precluded by roundoff errors.

Using the adjoint network method we can compute exact values of the group delay function. From (5.5) and $p = \omega$ we have

Table 5.1
Differential Scattering Matrices for Some Basic Elements of Microwave Networks

Circuit element	$\partial S/\partial p$	p
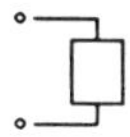 Z	$\dfrac{2Z_R}{Z + Z_R}$	Z
Z 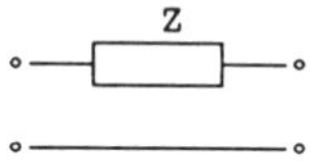	$\dfrac{2Z_R}{(Z + 2Z_R)^2}\begin{bmatrix} 1 & -1 \\ -1 & 1 \end{bmatrix}$	Z
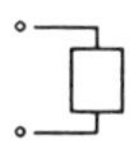 Y	$\dfrac{-2Z_R}{(1 + Z_R\, Y)^2}$	Y
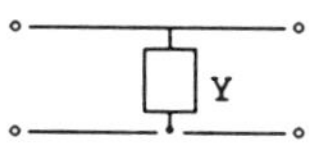	$\dfrac{-2Y_R}{(Y + 2Y_R)^2}\begin{bmatrix} 1 & 1 \\ 1 & 1 \end{bmatrix}$	Y
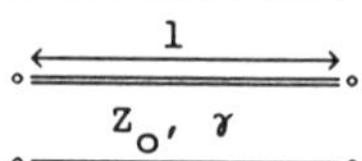	$\dfrac{1}{\left[Z_o^2 + Z_R^2 + 2\, Z_o\, Z_R \coth(\gamma l)\right]^2} \cdot \begin{bmatrix} 4Z_oZ_R^2+2Z_R(Z_o^2+Z_R^2)\coth(\gamma l) & 2Z_R(Z_R-Z_o^2)\operatorname{csch}(\gamma l) \\ 2Z_R(Z_R-Z_o^2)\operatorname{csch}(\gamma l) & 4Z_oZ_R^2+2Z_R(Z_o^2+Z_R^2)\coth(\gamma l) \end{bmatrix}$	Z_o
Transmission line section $\gamma = \alpha + j\beta \quad , \beta = \dfrac{2\Pi}{\lambda_g}$	$\dfrac{2Z_oZ_R\, \gamma \operatorname{csch}^2(\gamma l)}{\left[Z_o^2 + Z_R^2 + 2\, Z_o\, Z_R \coth(\gamma l)\right]^2} \cdot \begin{bmatrix} Z_o^2-Z_R^2 & -2Z_oZ_R\sinh(\gamma l) - (Z_o^2+Z_R^2)\cosh(\gamma l) \\ -2Z_oZ_R\sinh(\gamma l) - (Z_o^2+Z_R^2)\cosh(\gamma l) & Z_o^2-Z_R^2 \end{bmatrix}$	l

Table 5.1 (continued)

Circuit element	$\partial \underline{S}/\partial p$	p
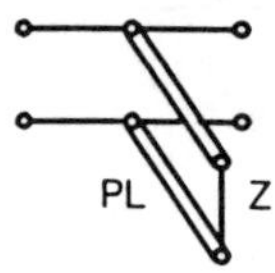 Short ended parallel stub	$\dfrac{-2\, Y_R \coth(\gamma l)}{\left[2\, Y_R + Y_o \coth(\gamma l)\right]^2} \cdot \begin{bmatrix} 1 & 1 \\ 1 & 1 \end{bmatrix}$	Y_o
	$\dfrac{2\, Y_R\, \gamma}{\sinh^2(\gamma l)\, \left[2\, Y_R + Y_o \coth(\gamma l)\right]^2} \cdot \begin{bmatrix} 1 & 1 \\ 1 & 1 \end{bmatrix}$	1
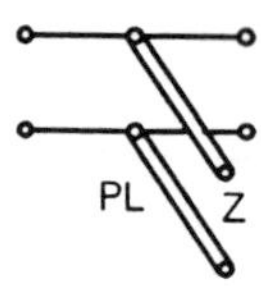 Open ended parallel stub	$\dfrac{-2\, Y_R \tanh(\gamma l)}{\left[2\, Y_R + Y_o \tanh(\gamma l)\right]^2} \cdot \begin{bmatrix} 1 & 1 \\ 1 & 1 \end{bmatrix}$	Y_o
	$\dfrac{2\, Y_R\, \gamma}{\cosh^2(\gamma l)\, \left[2\, Y_R + Y_o \tanh(\gamma l)\right]^2} \cdot \begin{bmatrix} 1 & 1 \\ 1 & 1 \end{bmatrix}$	1
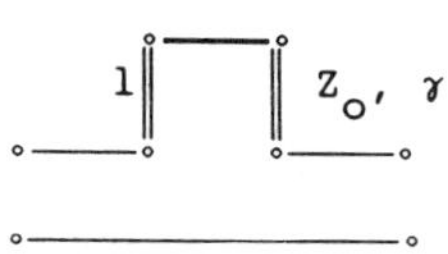 Short ended series stub	$\dfrac{2\, Z_R \tanh(\gamma l)}{\left[2\, Z_R + Z_o \tanh(\gamma l)\right]^2} \cdot \begin{bmatrix} 1 & -1 \\ -1 & 1 \end{bmatrix}$	Z_o
	$\dfrac{2\, Z_R\, \gamma}{\cosh^2(\gamma l)\, \left[2\, Z_R + Z_o \tanh(\gamma l)\right]^2} \cdot \begin{bmatrix} 1 & -1 \\ -1 & 1 \end{bmatrix}$	1
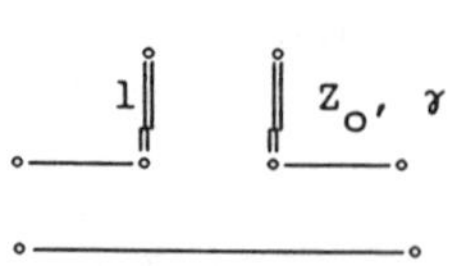 Open ended series stub	$\dfrac{2\, Z_R \coth(\gamma l)}{\left[2\, Z_R + Z_o \coth(\gamma l)\right]^2} \cdot \begin{bmatrix} 1 & -1 \\ -1 & 1 \end{bmatrix}$	Z_o
	$\dfrac{-\,2\, Z_R\, \gamma}{\sinh^2(\gamma l)\, \left[2\, Z_R + Z_o \coth(\gamma l)\right]^2} \cdot \begin{bmatrix} 1 & -1 \\ -1 & 1 \end{bmatrix}$	1

Table 5.1 (continued)

Circuit element	$\partial \underline{S}/\partial p$	p
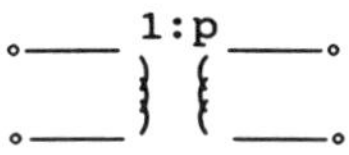 Ideal transformer	$\frac{2}{\left[1 + p^2\right]^2} \begin{bmatrix} -2p & 1-p^2 \\ 1-p^2 & -2p \end{bmatrix}$	p
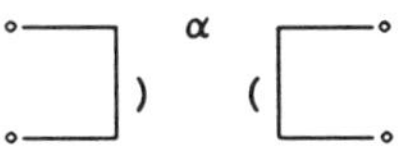 Gyrator	$\frac{1}{\left[\alpha^2 + Z_R^2\right]^2} \begin{bmatrix} 2\alpha\, Z_R & Z_R^2 - \alpha^2 \\ Z_R^2 - \alpha^2 & 2\alpha\, Z_R \end{bmatrix}$	α
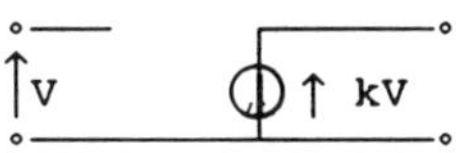 Voltage controlled voltage source	$\begin{bmatrix} 0 & 0 \\ 2 & 0 \end{bmatrix}$	k
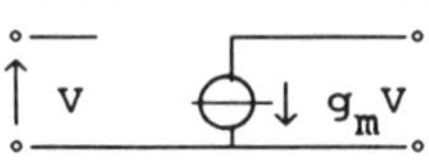 Voltage controlled current source	$\begin{bmatrix} 0 & 0 \\ -2Z_R & 0 \end{bmatrix}$	g_m
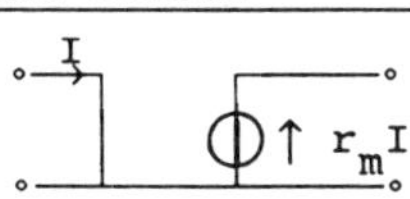 Current controlled voltage source	$\begin{bmatrix} 0 & 0 \\ \frac{2}{Z_R} & 0 \end{bmatrix}$	r_m
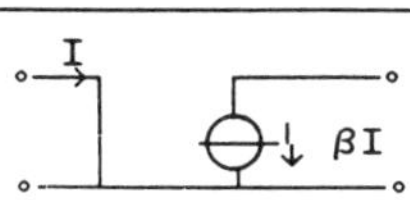 Current controlled current source	$\begin{bmatrix} 0 & 0 \\ -2 & 0 \end{bmatrix}$	β

$$\tau_{ij} = -\frac{\partial \phi_{ij}}{\partial \omega} = -\operatorname{Im}\left\{\frac{1}{S_{ij}}\frac{\partial S_{ij}}{\partial \omega}\right\} \tag{5.23}$$

Exciting the jth port of the circuit by a matched generator with $c = 1$ and terminating the kth output port with a matched load, we may compute the transmission function S_{ij} from

$$S_{ij}\Big|_{a_j=1} = b_i = a_l \tag{5.24}$$

where a_l is the ingoing wave variable at the lth port of the matched load terminating the ith port of the network.

To compute the partial derivative of S_{ij} with respect to ω, we refer to (5.18), in which we set $p = \omega$ [9, 10], and get

$$\frac{\partial S_{ij}}{\partial \omega} = \frac{\partial a_l}{\partial \omega} = \alpha^{\mathrm{T}}\frac{\partial \mathbf{S}}{\partial \omega}\mathbf{a} = \sum_{k \in E_\omega} \alpha^{(k)\mathrm{T}}\frac{\partial \mathbf{S}^{(k)}}{\partial \omega}\mathbf{a}^{(k)} \tag{5.25}$$

In the above relation E_ω represents a set of all frequency-dependent elements of the analyzed network.

Substituting (5.23) into (5.21) we get the relation for the exact computation of the group delay of any transmission function as

$$\tau_{ij} = -\operatorname{Im}\left\{\frac{1}{a_l}\sum_{k \in E_\omega} \alpha^{(k)\mathrm{T}}\frac{\partial \mathbf{S}^{(k)}}{\partial \omega}\mathbf{a}^{(k)}\right\} \tag{5.26}$$

Formulas for the right-hand side of (5.26) must be obtained for specific elements. To do this we may use the formulas presented in Table 5.2.

Example 5.1

Figure 5.2 presents a one-stage microwave transistor amplifier. The network consists of seven elements: a signal generator E1 and a load E7, two transmission line sections E3 and E5, two short-circuited shunt stubs E2 and E6, and a FET transistor E5. The transmission line sections together with short-circuited shunt stubs form matching networks at the input (E2 and E3) and at the output (E5 and E6) of the transistor.

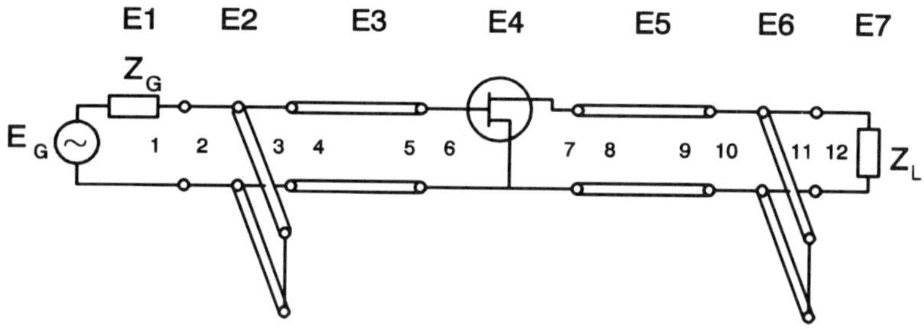

Figure 5.2 The network of a single-stage transistor amplifier composed of seven elements described by their own scattering matrices and noise wave correlation matrices.

The matrix equation $\mathbf{W}\,\mathbf{a} = \mathbf{c}$ of the circuit has the form

$$
\begin{array}{c}
 \\ 1 \\ 2 \\ 3 \\ 4 \\ 5 \\ 6 \\ 7 \\ 8 \\ 9 \\ 10 \\ 11 \\ 12
\end{array}
\begin{array}{c}
\begin{array}{cccccccccccc}
1 & 2 & 3 & 4 & 5 & 6 & 7 & 8 & 9 & 10 & 11 & 12
\end{array} \\
\left[\begin{array}{c:cc:cc:cc:cc:cc:c}
-\mathbf{S}^{(1)} & 1 & 0 & 0 & 0 & 0 & 0 & 0 & 0 & 0 & 0 & 0 \\ \hdashline
1 & \multicolumn{2}{c:}{-\mathbf{S}^{(2)}} & 0 & 0 & 0 & 0 & 0 & 0 & 0 & 0 & 0 \\
0 & & & 1 & 0 & 0 & 0 & 0 & 0 & 0 & 0 & 0 \\ \hdashline
0 & 0 & 1 & \multicolumn{2}{c:}{-\mathbf{S}^{(3)}} & 0 & 0 & 0 & 0 & 0 & 0 & 0 \\
0 & 0 & 0 & & & 1 & 0 & 0 & 0 & 0 & 0 & 0 \\ \hdashline
0 & 0 & 0 & 0 & 1 & \multicolumn{2}{c:}{-\mathbf{S}^{(4)}} & 0 & 0 & 0 & 0 & 0 \\
0 & 0 & 0 & 0 & 0 & & & 1 & 0 & 0 & 0 & 0 \\ \hdashline
0 & 0 & 0 & 0 & 0 & 0 & 1 & \multicolumn{2}{c:}{-\mathbf{S}^{(5)}} & 0 & 0 & 0 \\
0 & 0 & 0 & 0 & 0 & 0 & 0 & & & 1 & 0 & 0 \\ \hdashline
0 & 0 & 0 & 0 & 0 & 0 & 0 & 0 & 1 & \multicolumn{2}{c:}{-\mathbf{S}^{(6)}} & 0 \\
0 & 0 & 0 & 0 & 0 & 0 & 0 & 0 & 0 & & & 1 \\ \hdashline
0 & 0 & 0 & 0 & 0 & 0 & 0 & 0 & 0 & 0 & 1 & -\mathbf{S}^{(7)}
\end{array}\right]
\end{array}
\left[\begin{array}{c}
a_1 \\ a_2 \\ a_3 \\ a_4 \\ a_5 \\ a_6 \\ a_7 \\ a_8 \\ a_9 \\ a_{10} \\ a_{11} \\ a_{12}
\end{array}\right]
=
\left[\begin{array}{c}
1 \\ 0 \\ 0 \\ 0 \\ 0 \\ 0 \\ 0 \\ 0 \\ 0 \\ 0 \\ 0 \\ 0
\end{array}\right]
$$

where $\mathbf{S}^{(1)}, \mathbf{S}^{(2)}, \ldots, \mathbf{S}^{(7)}$ are scattering parameters of the seven circuit elements and are given in Table 2.1.

The fourth element is the FET transistor described by its scattering matrix $\mathbf{S}^{(4)}$. The scattering parameters of the transistor should be given as numerical values at a discrete set of frequency points at a given dc bias.

In the expressions defining the scattering matrices of particular circuit elements Z_R is the reference impedance common to all ports of the analyzed circuit. Because scattering parameters of FET transistors are usually determined in the $Z_O = 50\text{-}\Omega$ measurement systems, it is very convenient to assume that $Z_R = Z_O = 50\ \Omega$.

In the following discussion we assume that the overall $S_{21\text{ net}}$ parameter of the amplifier is the circuit function of interest. It will be computed according to the algorithm presented in Section 3.2.3 as

$$S_{21\text{ net}} = \left.\frac{a_L}{a_1}\right|_{a_1 = 1} = a_L = a_{12}$$

To evaluate the gradient vector $\nabla S_{21\text{ net}}$ of the overall $S_{21\text{ net}}$ parameter we must compute the sensitivities of the a_{12} wave variable, the ingoing wave variable at the load port (port number 12). To do this we solve a matrix equation $\mathbf{W}^T\ \alpha = \mathbf{e}_{12}$ of the adjoint network in the form given as (only the FET transistor is nonreciprocal two-port $(\mathbf{S}^{(4)T} \neq \mathbf{S}^{(4)})$

$$\begin{array}{c|cccccccccccc|}
 & 1 & 2 & 3 & 4 & 5 & 6 & 7 & 8 & 9 & 10 & 11 & 12 \\
\hline
1 & -\mathbf{S}^{(1)} & 1 & 0 & 0 & 0 & 0 & 0 & 0 & 0 & 0 & 0 & 0 \\
2 & 1 & \multicolumn{2}{c}{-\mathbf{S}^{(2)}} & 0 & 0 & 0 & 0 & 0 & 0 & 0 & 0 & 0 \\
3 & 0 & & & 1 & 0 & 0 & 0 & 0 & 0 & 0 & 0 & 0 \\
4 & 0 & 0 & 1 & \multicolumn{2}{c}{-\mathbf{S}^{(3)}} & 0 & 0 & 0 & 0 & 0 & 0 & 0 \\
5 & 0 & 0 & 0 & & & 1 & 0 & 0 & 0 & 0 & 0 & 0 \\
6 & 0 & 0 & 0 & 0 & 1 & \multicolumn{2}{c}{-\mathbf{S}^{(4)T}} & 0 & 0 & 0 & 0 & 0 \\
7 & 0 & 0 & 0 & 0 & 0 & & & 1 & 0 & 0 & 0 & 0 \\
8 & 0 & 0 & 0 & 0 & 0 & 0 & 1 & \multicolumn{2}{c}{-\mathbf{S}^{(5)}} & 0 & 0 & 0 \\
9 & 0 & 0 & 0 & 0 & 0 & 0 & 0 & & & 1 & 0 & 0 \\
10 & 0 & 0 & 0 & 0 & 0 & 0 & 0 & 0 & 1 & \multicolumn{2}{c}{-\mathbf{S}^{(6)}} & 0 \\
11 & 0 & 0 & 0 & 0 & 0 & 0 & 0 & 0 & 0 & & & 1 \\
12 & 0 & 0 & 0 & 0 & 0 & 0 & 0 & 0 & 0 & 0 & 1 & -\mathbf{S}^{(7)}
\end{array}
\begin{bmatrix} \alpha_1 \\ \alpha_2 \\ \alpha_3 \\ \alpha_4 \\ \alpha_5 \\ \alpha_6 \\ \alpha_7 \\ \alpha_8 \\ \alpha_9 \\ \alpha_{10} \\ \alpha_{11} \\ \alpha_{12} \end{bmatrix}
=
\begin{bmatrix} 0 \\ 0 \\ 0 \\ 0 \\ 0 \\ 0 \\ 0 \\ 0 \\ 0 \\ 0 \\ 0 \\ 1 \end{bmatrix}$$

The sensitivity of the wave variable a_{12} with respect to any variable network parameter is computed using (5.11) and formulas given in Table 5.1.

We assume now that the variable parameters of the circuit are Y_{O2}, Z_{O2}, Z_{O5}, Y_{O6}, l_2, l_3, l_5, and l_6. The gradient operator is then given as

$$\nabla = \left[\frac{\partial}{\partial Y_{O2}}, \frac{\partial}{\partial Z_{O3}}, \frac{\partial}{\partial Z_{O5}}, \frac{\partial}{\partial Y_{O6}}, \frac{\partial}{\partial l_2}, \frac{\partial}{\partial l_3}, \frac{\partial}{\partial l_5}, \frac{\partial}{\partial l_6}\right]^T$$

The sensitivity of the wave variable a_{12} with respect to the characteristic impedance Z_O of a transmission line section (elements E3 and E5) is equal to

$$\frac{\partial a_{12}}{\partial Z_O} = \alpha^T \frac{\partial \mathbf{S}}{\partial Z_O} \mathbf{a}$$

Then we have

$$G_2 = \frac{\partial a_{12}}{\partial Z_{O3}} = \frac{1}{[Z_{O3}^2 + Z_R^2 + 2\, Z_{O3}\, Z_R \coth(\gamma l_3)]^2}$$

$$\cdot \begin{bmatrix} \alpha_4 \\ \alpha_5 \end{bmatrix}^T \begin{bmatrix} 4Z_{O3}Z_R^2 + 2Z_R\,(Z_{O3}^2 + Z_R^2)\coth(\gamma l_3) & 4Z_{O3}Z_R^2 + 2Z_R\,(Z_{O3}^2 + Z_R^2)\coth(\gamma l_3) \\ 2Z_R\,(Z_R - Z_{O3}^2)\operatorname{csch}(\gamma l_3) & 2Z_R\,(Z_R - Z_{O3}^2)\operatorname{csch}(\gamma l_3) \end{bmatrix} \begin{bmatrix} a_4 \\ a_5 \end{bmatrix}$$

$$G_3 = \frac{\partial a_{12}}{\partial Z_{O5}} = \frac{1}{[Z_{O5}^2 + Z_R^2 + 2\, Z_{O5}\, Z_R \coth(\gamma l_5)]^2}$$

$$\cdot \begin{bmatrix} \alpha_8 \\ \alpha_9 \end{bmatrix}^T \begin{bmatrix} 4Z_{O5}Z_R^2 + 2Z_R\,(Z_{O5}^2 + Z_R^2)\coth(\gamma l_5) & 4Z_{O5}Z_R^2 + 2Z_R\,(Z_{O5}^2 + Z_R^2)\coth(\gamma l_5) \\ 2Z_R\,(Z_R - Z_{O5}^2)\operatorname{csch}(\gamma l_5) & 2Z_R\,(Z_R - Z_{O5}^2)\operatorname{csch}(\gamma l_5) \end{bmatrix} \begin{bmatrix} a_8 \\ a_9 \end{bmatrix}$$

The sensitivity of the wave variable a_{12} with respect to the length l of a transmission line section (elements E3 and E5) is

$$\frac{\partial a_{12}}{\partial l} = \alpha^T \frac{\partial \mathbf{S}}{\partial l} \mathbf{a}$$

According to the above relation we have

$$G_6 = \frac{\partial a_{12}}{\partial l_3} = \frac{2Z_{O3}Z_R\,\gamma \operatorname{csch}^2(\gamma l_3)}{[Z_{O3}^2 + Z_R^2 + 2\, Z_{O3}Z_R \coth(\gamma l_3)]^2}$$

$$\cdot \begin{bmatrix} \alpha_4 \\ \alpha_5 \end{bmatrix}^{\mathrm{T}} \begin{bmatrix} Z_{\mathrm{O3}}^2 - Z_{\mathrm{R}}^2 & -2Z_{\mathrm{O3}}Z_{\mathrm{R}} \sinh(\gamma l_3) - (Z_{\mathrm{O3}}^2 + Z_{\mathrm{R}}^2) \cosh(\gamma l_3) \\ -2Z_{\mathrm{O3}}Z_{\mathrm{R}} \sinh(\gamma l) - (Z_{\mathrm{O3}}^2 + Z_{\mathrm{R}}^2) \cosh(\gamma l_3) & Z_{\mathrm{O3}}^2 - Z_{\mathrm{R}}^2 \end{bmatrix} \begin{bmatrix} a_4 \\ a_5 \end{bmatrix}$$

$$G_7 = \frac{\partial a_{12}}{\partial l_5} = \frac{2Z_{\mathrm{O3}}Z_{\mathrm{R}}\, \gamma \operatorname{csch}^2(\gamma l_3)}{[Z_{\mathrm{O3}}^2 + Z_{\mathrm{R}}^2 + 2\, Z_{\mathrm{O3}}Z_{\mathrm{R}} \coth(\gamma l_3)]^2}$$

$$\cdot \begin{bmatrix} \alpha_8 \\ \alpha_9 \end{bmatrix}^{\mathrm{T}} \begin{bmatrix} Z_{\mathrm{O5}}^2 - Z_{\mathrm{R}}^2 & -2Z_{\mathrm{O5}}Z_{\mathrm{R}} \sinh(\gamma l_5) - (Z_{\mathrm{O5}}^2 + Z_{\mathrm{R}}^2) \cosh(\gamma l_5) \\ -2Z_{\mathrm{O5}}Z_{\mathrm{R}} \sinh(\gamma l) - (Z_{\mathrm{O5}}^2 + Z_{\mathrm{R}}^2) \cosh(\gamma l_5) & Z_{\mathrm{O5}}^2 - Z_{\mathrm{R}}^2 \end{bmatrix} \begin{bmatrix} a_8 \\ a_9 \end{bmatrix}$$

Sensitivity of the wave variable a_{12} with respect to the characteristic admittance of a short-circuited parallel stub is

$$\frac{\partial a_{12}}{\partial Y_{\mathrm{O}}} = \alpha^{\mathrm{T}} \frac{\partial \mathbf{S}}{\partial Y_{\mathrm{O}}} \mathbf{a}$$

Sensitivity of the wave variable a_{12} with respect to the length l of a parallel stub is

$$\frac{\partial a_{12}}{\partial l} = \alpha^{\mathrm{T}} \frac{\partial \mathbf{S}}{\partial l} \mathbf{a}$$

Then, we have

$$G_1 = \frac{\partial a_{12}}{\partial Y_{\mathrm{O2}}} = \frac{-2\, Y_{\mathrm{R}} \coth(\gamma l_2)}{[2\, Y_{\mathrm{R}} + Y_{\mathrm{O2}} \coth(\gamma l_2)]^2} \cdot \begin{bmatrix} \alpha_2 \\ \alpha_3 \end{bmatrix}^{\mathrm{T}} \begin{bmatrix} 1 & 1 \\ 1 & 1 \end{bmatrix} \begin{bmatrix} a_2 \\ a_3 \end{bmatrix}$$

$$G_4 = \frac{\partial a_{12}}{\partial Y_{\mathrm{O6}}} = \frac{-2\, Y_{\mathrm{R}} \coth(\gamma l_6)}{[2\, Y_{\mathrm{R}} + Y_{\mathrm{O6}} \coth(\gamma l_6)]^2} \cdot \begin{bmatrix} \alpha_{10} \\ \alpha_{11} \end{bmatrix}^{\mathrm{T}} \begin{bmatrix} 1 & 1 \\ 1 & 1 \end{bmatrix} \begin{bmatrix} a_{10} \\ a_{11} \end{bmatrix}$$

$$G_5 = \frac{\partial a_{12}}{\partial l_2} = \frac{2\, Y_{\mathrm{R}}\gamma}{\sinh^2(\gamma l_2)\, [2\, Y_{\mathrm{R}} + Y_{\mathrm{O2}} \coth(\gamma l_2)]^2} \cdot \begin{bmatrix} \alpha_2 \\ \alpha_3 \end{bmatrix}^{\mathrm{T}} \begin{bmatrix} 1 & 1 \\ 1 & 1 \end{bmatrix} \begin{bmatrix} a_2 \\ a_3 \end{bmatrix}$$

$$G_8 = \frac{\partial a_{12}}{\partial l_6} = \frac{2\, Y_{\mathrm{R}}\gamma}{\sinh^2(\gamma l_6)\, [2\, Y_{\mathrm{R}} + Y_{\mathrm{O6}} \coth(\gamma l_6)]^{\mathrm{T}}} \cdot \begin{bmatrix} \alpha_{10} \\ \alpha_{11} \end{bmatrix}^{\mathrm{T}} \begin{bmatrix} 1 & 1 \\ 1 & 1 \end{bmatrix} \begin{bmatrix} a_{10} \\ a_{11} \end{bmatrix}$$

The gradient vector of the $S_{21\ \text{net}}$ is equal to

$$\nabla S_{21\ \text{net}} = \mathbf{G} = [G_1, G_2, G_3, G_4, G_5, G_6]^{\text{T}}$$

All elements of **G** are complex quantities and may be split into two parts: the sensitivity of the magnitude of the $S_{21\ \text{net}}$ parameter and the sensitivity of the argument Φ of the $S_{21\ \text{net}}$ parameter. According to (5.4) and (5.5) we have

$$\nabla |S_{21\ \text{net}}| = |S_{21\ \text{net}}| \cdot \text{Re}\left\{\frac{1}{S_{21\ \text{net}}}\begin{bmatrix} G_1 \\ G_2 \\ G_3 \\ G_4 \\ G_5 \\ G_6 \end{bmatrix}\right\}$$

$$\nabla \arg\{S_{21\ \text{net}}\} = \text{Im}\left\{\frac{1}{S_{21\ \text{net}}}\begin{bmatrix} G_1 \\ G_2 \\ G_3 \\ G_4 \\ G_5 \\ G_6 \end{bmatrix}\right\}$$

5.4 NOISE POWER SENSITIVITY ANALYSIS

Suppose that a parameter p in a given noisy network is to be varied without affecting its topology. Parameter variation will affect the noise power delivered to the load at the rth port of the network. Differentiating the equation

$$N_l = (\overline{\mathbf{a}\,\mathbf{a}^+})_{ll} = [(\mathbf{W}^+)^{-1}\,\mathbf{e}_l]^+\,\mathbf{C}(\mathbf{W}^+)^{-1}\,\mathbf{e}_l \tag{5.27}$$

with respect to p and using the relation $\partial\mathbf{W}/\partial p = -\partial\mathbf{S}/\partial p$, we have [11, 12]

$$\frac{\partial N_l}{\partial p} = \frac{\partial(\overline{\mathbf{a}\,\mathbf{a}^+})_{ll}}{\partial p} = \mathbf{e}_l^+\,\mathbf{W}^{-1}\left[\frac{\partial\mathbf{S}}{\partial p}\mathbf{W}^{-1}\,\mathbf{C} + \left(\frac{\partial\mathbf{S}}{\partial p}\mathbf{W}^{-1}\,\mathbf{C}^+\right)^+\right](\mathbf{W}^+)^{-1}\mathbf{e}_l$$

$$
+ e_l^{+} \mathbf{W}^{-1} \frac{\partial \mathbf{C}}{\partial p} (\mathbf{E}^{+})^{-1} \mathbf{e}_l
$$

$$
= \underline{\mathcal{A}}^{+} \left[\frac{\partial \mathbf{S}}{\partial p} \mathbf{W}^{-1} \mathbf{C} + \left(\frac{\partial \mathbf{S}}{\partial p} \mathbf{W}^{-1} \mathbf{C}^{+} \right)^{+} \right] \underline{\mathcal{A}}
$$

$$
+ \underline{\mathcal{A}}^{+} \frac{\partial \mathbf{C}}{\partial p} \underline{\mathcal{A}} \tag{5.28}
$$

Because noise correlation matrices are Hermitian matrices ($\mathbf{C} = \mathbf{C}^{+}$), (5.28) simplifies to

$$
\frac{\partial N_l}{\partial p} = 2 \operatorname{Re} \left\{ \underline{\mathcal{A}}^{+} \frac{\partial \mathbf{S}}{\partial p} \mathbf{W}^{-1} \mathbf{C} \underline{\mathcal{A}} \right\} + \underline{\mathcal{A}}^{+} \frac{\partial \mathbf{C}}{\partial p} \underline{\mathcal{A}} \tag{5.29}
$$

Further, assuming that a varied parameter p is contained in the passive part of the circuit, we have

$$
\frac{\partial N_l}{\partial p} = 2 \operatorname{Re} \left\{ \underline{\mathcal{A}}^{+} \frac{\partial \mathbf{S}}{\partial p} \mathbf{W}^{-1} \mathbf{C} \underline{\mathcal{A}} \right\} - 2 \underline{\mathcal{A}}^{+} \operatorname{Re} \left\{ \frac{\partial \mathbf{S}}{\partial p} \mathbf{S}^{+} \right\} \underline{\mathcal{A}} \tag{5.30}
$$

Using (5.30), we may write

$$
\mathbf{G} = \begin{bmatrix} \dfrac{\partial N_l}{\partial p_1} \\ \dfrac{\partial N_l}{\partial p_2} \\ \vdots \\ \dfrac{\partial N_l}{\partial p_n} \end{bmatrix} = 2 \operatorname{Re} \left\{ \underline{\mathcal{A}}^{+} \begin{bmatrix} \dfrac{\partial \mathbf{S}}{\partial p_1} \\ \dfrac{\partial \mathbf{S}}{\partial p_2} \\ \vdots \\ \dfrac{\partial \mathbf{S}}{\partial p_n} \end{bmatrix} \mathbf{W}^{-1} \mathbf{C} \underline{\mathcal{A}} \right\} - 2 \underline{\mathcal{A}}^{+} \operatorname{Re} \left\{ \begin{bmatrix} \dfrac{\partial \mathbf{S}}{\partial p_1} \\ \dfrac{\partial \mathbf{S}}{\partial p_2} \\ \vdots \\ \dfrac{\partial \mathbf{S}}{\partial p_n} \end{bmatrix} \mathbf{S}^{+} \right\} \underline{\mathcal{A}} \tag{5.31}
$$

where $\mathbf{G} = \nabla N_l$ is a vector of sensitivities of noise power dissipated in the load located in the rth port of the analyzed circuit.

Equation (5.30) or (5.31) relates changes in noise power dissipated in the load to changes in parameter values in the passive elements of the network.

To evaluate the noise power sensitivities, we need to know the partial derivatives of element scattering matrices, the $\underline{\mathcal{A}}$ vector being the solution of the system of equations

$$\mathbf{W}^{\mathrm{T}*} \underline{\mathcal{A}} = \mathbf{e}_l \tag{5.32}$$

and the inverse of the connection scattering matrix of the analyzed network.

The two first quantities are used also in sensitivity analysis of networks excited by the sinusoidal signals discussed in Section 5.1. The sensitivity of sinusoidal wave a_r at the rth port is obtained from (5.12) as

$$\frac{\partial a_r}{\partial p} = \alpha^{\mathrm{T}} \frac{\partial \mathbf{S}}{\partial p} \mathbf{a} \tag{5.33}$$

where **a** is a solution vector of the system of equations

$$\mathbf{W}\, \mathbf{a} = \mathbf{c} \tag{5.34}$$

of the original circuit and α is a solution vector of a system of equations

$$\mathbf{W}^{\mathrm{T}}\, \alpha = \mathbf{e}_r \tag{5.35}$$

of the adjoined circuit, in which $\mathbf{e}_r$ is a vector of the same form as the vector $\mathbf{e}_l$ given by (4.125).

Comparing (5.32) and (5.35) we find that the solution vectors of both systems of equations satisfy the relation

$$\underline{\mathcal{A}} = \alpha^* \tag{5.36}$$

which means that to have $\underline{\mathcal{A}}$ we must find α or vice versa. This significantly reduces computational effort when performing noise analysis, noise power sensitivity analysis, and sensitivity analysis for deterministic signals at the same time.

Partial derivatives of the element scattering matrices required for evaluation of noise power sensitivities can usually be found after some manipulations. Table 5.1 presents scattering matrix partial derivatives for some basic microwave network elements with respect to their useful parameters.

The inverse matrices $\mathbf{W}^{-1}$ and $(\mathbf{W}^{-1})^+$ of the connection scattering matrix $\mathbf{W}$ are computed very effectively using a sparse matrix technique based on the bi-factorization method that is discussed in Chapter 6.

5.5 NOISE FIGURE GRADIENT VECTOR COMPUTATION

The noise figure of a network can be computed from relations (4.135), (4.136), and (4.131) as

$$F = 1 + \frac{N_{\text{int}\ l}}{N_{\text{g}\ l}} \tag{5.37}$$

where

$$N_{\text{int}\ l} = \underline{\mathscr{A}}^+ \, \mathbf{C}_{\text{int}} \, \underline{\mathscr{A}} \tag{5.38}$$

and

$$N_{\text{G}\ l} = \underline{\mathscr{A}}^+ \, \mathbf{C}_{\text{G}} \, \underline{\mathscr{A}} \tag{5.39}$$

Differentiating the relation (5.37) with respect to a circuit parameter p, using relations (5.38), (5.39), and relation (5.29) describing the noise power sensitivities, we get the expression for noise figure sensitivity relating to a parameter p [12] as

$$\begin{aligned} \frac{\partial F}{\partial p} &= \frac{1}{N_{\text{G}\ l}^2}\left(\frac{\partial N_{\text{int}\ l}}{\partial p} N_{\text{G}\ l} - \frac{\partial N_{\text{G}\ l}}{\partial p} N_{\text{int}\ l}\right) \\ &= \frac{1}{\underline{\mathscr{A}}^+ \mathbf{C}_{\text{G}} \underline{\mathscr{A}}}\left[2\,\text{Re}\left\{\underline{\mathscr{A}}^+ \frac{\partial \mathbf{S}}{\partial p} \mathbf{W}^{-1} \, \mathbf{C}_{\text{int}} \, \underline{\mathscr{A}}\right\} + \underline{\mathscr{A}}^+ \frac{\partial \mathbf{C}_{\text{int}}}{\partial p} \underline{\mathscr{A}}\right] \\ &\quad - \frac{\underline{\mathscr{A}}^+ \, \mathbf{C}_{\text{int}} \, \underline{\mathscr{A}}}{(\underline{\mathscr{A}}^+ \mathbf{C}_{\text{G}} \underline{\mathscr{A}})^2}\, 2\,\text{Re}\left\{\underline{\mathscr{A}}^+ \frac{\partial \mathbf{S}}{\partial p} \mathbf{W}^{-1} \, \mathbf{C}_{\text{G}} \, \underline{\mathscr{A}}\right\} \end{aligned} \tag{5.40}$$

In deriving (5.40) we assumed that the thermal noise generated by the internal impedance of a signal generator did not depend on the varied circuit parameter p.

Now consider a design of a circuit. Suppose it is desired to optimize (minimize) the noise figure of a circuit. Expending the relation (5.40) we have a gradient vector of the noise figure F considered as an objective function [12]:

$$\nabla F = \begin{bmatrix} \dfrac{\partial F}{\partial p_1} \\ \dfrac{\partial F}{\partial p_2} \\ \vdots \\ \dfrac{\partial F}{\partial p_n} \end{bmatrix} = \frac{1}{\underline{\mathscr{A}}^+ \mathbf{C}_G \underline{\mathscr{A}}} \left[2\,\mathrm{Re}\left\{ \underline{\mathscr{A}}^+ \begin{bmatrix} \dfrac{\partial \mathbf{S}}{\partial p_1} \\ \dfrac{\partial \mathbf{S}}{\partial p_2} \\ \vdots \\ \dfrac{\partial \mathbf{S}}{\partial p_n} \end{bmatrix} \mathbf{W}^{-1} \mathbf{C}_{\mathrm{int}} \underline{\mathscr{A}} \right\} + \underline{\mathscr{A}}^+ \begin{bmatrix} \dfrac{\partial \mathbf{C}_{\mathrm{int}}}{\partial p_1} \\ \dfrac{\partial \mathbf{C}_{\mathrm{int}}}{\partial p_2} \\ \vdots \\ \dfrac{\partial \mathbf{C}_{\mathrm{int}}}{\partial p_n} \end{bmatrix} \underline{\mathscr{A}} \right]$$

$$- \frac{\underline{\mathscr{A}}^+ \mathbf{C}_{\mathrm{int}} \underline{\mathscr{A}}}{(\underline{\mathscr{A}}^+ \underline{\mathbf{C}}_G \underline{\mathscr{A}})^2} 2\,\mathrm{Re}\left\{ \underline{\mathscr{A}}^+ \begin{bmatrix} \dfrac{\partial \mathbf{S}}{\partial p_1} \\ \dfrac{\partial \mathbf{S}}{\partial p_2} \\ \vdots \\ \dfrac{\partial \mathbf{S}}{\partial p_n} \end{bmatrix} \underline{W}^{-1} \underline{\mathscr{C}}_G \underline{\mathscr{A}} \right\} \tag{5.41}$$

Equation (5.41) is a general one. It is applicable to cases when variable parameters are contained in passive as well as active elements of a circuit.

References

[1] Brayton, R. K., and R. Spence, *Sensitivity and Optimization,* New York: Elsevier Scientific Publishing Company, 1980.

[2] Spence, R., and R. S. Soin, *Tolerance Design of Electronic Circuits,* Wokingham, England: Addison Wesley Publishing Company, 1988.

[3] Meehan, M. D., and J. Purviance, *Yield and Reliability in Microwave Circuit and System Design,* Norwood, MA: Artech House, 1993.

[4] Dobrowolski, J. A., *Introduction to Computer Methods for Microwave Circuit Analysis and Design,* Norwood, MA: Artech House, 1991.

[5] Iuculano, G., V. A. Monaco, and P. Tiberio, "Network Sensitivities in Terms of Scattering Parameters," *Electronic Lett.,* Vol. 8, 1972, pp. 53–54.

[6] Monaco, V. A., and P. Tiberio, "Two Properties for Circuit Sensitivities in Terms of Scattering Parameters," *Electronic Lett.,* Vol. 8, 1972, pp. 382–383.

[7] Bandler, J. W., and R. E. Seviora, "Wave Sensitivities of Networks," *IEEE Trans. Microwave Theory Tech.,* Vol. MTT-20, 1972, pp. 138–147.

[8] Monaco, V. A., and P. Tiberio, "On Linear Network Scattering Matrix Sensitivities," *Alta Frequenza,* Vol. 39, 1970, pp. 193–195.

[9] Monaco, V. A., and P. Tiberio, "Computer-Aided Analysis of Microwave Circuits," *IEEE Trans. Microwave Theory Tech.,* Vol. MTT-22, 1974, pp. 249–263.

[10] Iuculano, G., V. A. Monaco, and P. Tiberio, "A Computer Program for Sensitivity and Group Delay Evaluation of Linear Networks," *Alta Frequenza,* Vol. 40, No. 11, 1971, pp. 873–880.

[11] Dobrowolski, J. A., "A CAD Oriented Method for Noise Figure Computation of Two-Ports with any Internal Topology," *IEEE Trans. Microwave Theory Tech.,* Vol. MTT-37, January 1989, pp. 15–20.

[12] Dobrowolski, J. A., "Noise Power Sensitivities and Noise Figure Minimization of Two-Ports with Any Internal Topology," *IEEE Trans. Microwave Theory Tech.,* Vol. MTT-39, January 1991, pp. 136–140.

CHAPTER 6

Solution of Systems of Circuit Equations Using Sparse Matrix Techniques

It is clear from Chapters 3, 4, and 5 that many of the problems found in microwave circuit simulation can be ultimately reduced to solving related systems of linear algebraic equations. The matrix equations used in the analysis of microwave circuits using the connection scattering matrix approach are

$$\mathbf{W}\,\mathbf{a} = \mathbf{c} \tag{6.1}$$

$$\mathbf{W}^{\mathrm{T}}\,\alpha = \mathbf{e}_r \tag{6.2}$$

$$\mathbf{W}^{+}\,\underline{\mathcal{A}} = \mathbf{e}_l \tag{6.3}$$

where $\mathbf{W}$ is the $n \times n$ matrix of complex coefficients.

Equation (6.1) corresponds to the frequency domain analysis, whereas (6.2) is related to the sensitivity analysis of microwave networks. The last matrix equation (6.3) is used in the computer-aided noise analysis of microwave circuits.

In the computer-aided procedures of the analysis and design these equations must be solved many times. It may be shown that the total number of multiplications and divisions required for the solution of a system of linear equations is [1–5]

$$N_t = \frac{n^3}{3} + n^2 - \frac{n}{3} \tag{6.4}$$

where n is the order of the coefficient matrix.

It must be noted that the operation count as given above is for a system $\mathbf{A}\,\mathbf{x} = \mathbf{b}$ in which $\mathbf{A}$ is assumed to be full.

The full matrix also requires storage for $n^2 + n$ complex numbers. As it was shown in Chapter 3, the connection scattering matrix **W** is sparsely filled with entries of nonzero values.

Special techniques devised to take advantage of the sparsity that occurs in circuit equations are sparse matrix techniques. In effect, they reduce both the number of arithmetic operations required to compute the solution and the storage requirements. The amount of reduction is often quite dramatic. It depends on the sparsity of the matrix and the ingenuity of the programmer.

Connection scattering matrices of real microwave networks are very sparse, and it is advantageous to exploit their sparsity even when the number of equations is small. Exploiting sparsity of the circuit matrices allows us to reduce drastically the required execution time. It is typical of sparse matrix techniques that the number of multiplications and divisions required to solve the system grows approximately linearly with the size n, as compared to n^3 (and n^2) for full matrices. By storing only nonzero elements of the matrix, the amount of computer memory required is also drastically reduced.

There are three basic ideas used in sparse matrix techniques. The first is not to store matrix entries that have zero values because, first and foremost, the required amount of computer memory for matrix data storage is reduced. In addition, because the zeros are not stored, there is no need to access them, hence reducing execution time.

The second goal in the sparse matrix techniques is to perform only nontrivial arithmetic operations with nonzero elements. Because zero elements are not stored, multiplications by zero and additions with zero-valued elements may be eliminated by using appropriate programming techniques.

In order to exploit the sparsity of a matrix, the sparsity must not be destroyed during the decomposition procedure. The number of nonzero elements in the matrix may grow during its reduction. The new elements that come into existence in empty places of a matrix are called *fill-ins*. The growth of the number of fill-ins can be minimized by proper pivoting strategy. Maintaining sparsity during the reduction of the matrix is essential.

The technique described in this chapter is a sparse matrix technique based on the bi-factorization method with the application of a looping indexed code approach. The equation solver takes adventage of the characteristics of the connection scattering matrix.

6.1 BI-FACTORIZATION METHOD

In the bi-factorization method the inverse of the coefficient matrix **A** in the system of linear equations (6.1) is expressed as a multiple product of $2n$ factor matrices [6] as

$$\mathbf{A}^{-1} = \mathbf{R}^{(1)}\mathbf{R}^{(2)} \cdots \mathbf{R}^{(n-1)}\mathbf{R}^{(n)}\mathbf{L}^{(n)}\mathbf{L}^{(n-1)} \cdots \mathbf{L}^{(2)}\mathbf{L}^{(1)} \tag{6.5}$$

where $\mathbf{L}^{(i)}$, $i = 1, 2, \ldots, n$, are the left-hand factor matrices and $\mathbf{R}^{(i)}$, $i = 1, 2, \ldots, n$, are right-hand factor matrices.

The left-hand factor matrices $\mathbf{L}^{(k)}$ differ from the unity matrix only in the kth column:

$$\mathbf{L}^{(k)} = \begin{bmatrix} 1 & 0 & . & . & 0 & \cdots\cdots & 0 \\ 0 & 1 & . & & \vdots & & \\ \vdots & & & . & l_{kk}^{(k)} & & \\ \vdots & & & & l_{k+1,k}^{(k)} & 1 & \\ \vdots & & & & \vdots & & \ddots \\ 0 & . & . & . & l_{n,k}^{(k)} & \cdots\cdots & 1 \end{bmatrix} \tag{6.6}$$

Note that the diagonal term $l_{kk}^{(k)}$ of $\mathbf{L}^{(k)}$ is not equal to 1.

The right-hand factor matrices $\mathbf{R}^{(k)}$ differ from the unity matrix only in the kth row:

$$\mathbf{R}^{(k)} = \begin{bmatrix} 1 & 0 & \cdots & & & & & 0 \\ 0 & . & & & & & & \\ & & . & & & & & \\ & & & 1 & & & & \\ 0 & 0 & . & 0 & 1 & r_{k,k+1}^{(k)} & r_{k,k+2}^{(k)} \cdots & r_{k,n}^{(k)} \\ . & & & & & 1 & & 0 \\ . & & & & & & . & . \\ . & & & & & & . & . \\ . & & & & & & . & . \\ . & & & & & & 1 & 0 \\ 0 & 0 & \cdots & & & & 0 & 1 \end{bmatrix} \tag{6.7}$$

In $\mathbf{R}^{(k)}$ all diagonal elements are equal to 1 and thus $\mathbf{R}^{(n)} = \mathbf{I}$.

The factor matrices $\mathbf{L}^{(i)}$ and $\mathbf{R}^{(i)}$, $i = 1, 2, \ldots, n$, can be found from the equation

$$\mathbf{L}^{(n)}\mathbf{L}^{(n-1)} \cdots \mathbf{L}^{(2)}\mathbf{L}^{(1)}\mathbf{A}\,\mathbf{R}^{(1)}\mathbf{R}^{(2)} \cdots \mathbf{R}^{(n-1)}\mathbf{R}^{(n)} = \mathbf{I} \tag{6.8}$$

which can be written as a sequence of the intermediate matrices

$$
\begin{aligned}
\mathbf{A}^{(0)} &= \mathbf{A} \\
\mathbf{A}^{(1)} &= \mathbf{L}^{(1)}\, \mathbf{A}^{(0)}\, \mathbf{L}^{(1)} \\
\mathbf{A}^{(2)} &= \mathbf{L}^{(2)}\, \mathbf{A}^{(1)}\, \mathbf{L}^{(2)} \\
&\vdots \\
\mathbf{A}^{(k)} &= \mathbf{L}^{(k)}\, \mathbf{A}^{(k-1)}\, \mathbf{L}^{(k)} \\
&\vdots \\
\mathbf{A}^{(n)} &= \mathbf{L}^{(n)}\, \mathbf{A}^{(n-1)}\, \mathbf{L}^{(n)}
\end{aligned}
\tag{6.9}
$$

The above equations represent the step-by-step transformation of the initial matrix $\mathbf{A} = \mathbf{A}^{(0)}$ to the identity matrix $\mathbf{I}$. This is achieved by forming the successive inner products $\mathbf{L}^{(k)}\mathbf{A}^{(k-1)}\mathbf{R}^{(k)}$ for $k = 1, 2, \ldots, n$.

Using (6.5) to (6.9) it can be found that the elements of the reduced matrix $\mathbf{A}^{(k)}$ are computed as

$$
a_{kk}^{(k)} = 1 \qquad a_{kj}^{(k)} = 0 \qquad a_{ik}^{(k)} = 0 \tag{6.10}
$$

$$
a_{ij}^{(k)} = a_{ij}^{(k-1)} - \frac{a_{ik}^{(k-1)}\, a_{kj}^{(k-1)}}{a_{kk}^{(k-1)}} \tag{6.11}
$$

where k is the pivotal index and $i, j = (k+1), \ldots, n$.

The elements of the left-hand factor matrices are

$$
l_{kk}^{(k)} = \frac{1}{a_{kk}^{(k-1)}} \tag{6.12}
$$

$$
l_{ik}^{(k)} = -\frac{a_{ik}^{(k-1)}}{a_{kk}^{(k-1)}} \qquad i = (k+1), \ldots, n. \tag{6.13}
$$

In a similar way we compute the elements of the right-hand side factor matrices as

$$
r_{kk}^{(k)} = 1 \tag{6.14}
$$

$$
r_{kj}^{(k)} = -\frac{a_{kj}^{(k-1)}}{a_{kk}^{(k-1)}} \qquad j = (k+1), \ldots, n. \tag{6.15}
$$

As in the case of other methods, the pivots $a_{kk}^{(k-1)}$, $k = 1, 2, \ldots, n$, must be different from zero.

The vector of unknowns, that is, the solution vector, may be computed from the relation

$$\mathbf{x} = \mathbf{A}^{-1}\mathbf{b} = \mathbf{R}^{(1)}\mathbf{R}^{(2)} \cdots \mathbf{R}^{(n-1)}\mathbf{R}^{(n)}\mathbf{L}^{(n)}\mathbf{L}^{(n-1)} \cdots \mathbf{L}^{(2)}\mathbf{L}^{(1)}\mathbf{b} \qquad (6.16)$$

According to the above relation the right-hand side vector **b** is stepwise transformed to the solution vector by successive factor matrices by vector multiplications.

The computer implementation of the algorithm can be written directly from (6.10) to (6.16) as follows.

6.1.1 Bi-Factorization

Step 0: Set $k = 1$.
Step 1: Compute the kth column of $\mathbf{L}^{(k)}$ by setting

$$l_{kk}^{(k)} = \frac{1}{a_{kk}^{(k-1)}}$$

$$l_{ik}^{(k)} = -a_{ik}^{(k-1)}\, l_{kk}^{(k)} \qquad i = (k+1), \ldots, n$$

where $a_{ij}^{(k-1)}$ are entries of the reduced matrix $\mathbf{A}^{(k-1)}$ and $\mathbf{A}^{(0)} = \mathbf{A}$.
Step 2: If $k = n$, then stop.
Step 3: Update the reduced matrix from $\mathbf{A}^{(k-1)}$ to $\mathbf{A}^{(k)}$ by setting

$$a_{ij}^{(k)} = a_{ij}^{(k-1)} + a_{kj}^{(k-1)}\, l_{ik}^{(k)}$$

where $i, j = k+1, k+2, \ldots, n$.
Step 4: Obtain the kth row of $\mathbf{R}^{(k)}$ by setting

$$r_{kj}^{(k)} = -a_{kj}^{(k-1)}\, l_{kk}^{(k)}$$

Step 5: $k = k + 1$ and go to Step 1.

6.1.2 Computation of the Solution Vector

Set the solution vector to the right-hand side vector

$$x_i = b_i \qquad i = 1, 2, \ldots, n$$

The left-hand factor matrix by vector multiplications;

Step 0: Set $k = 1$.

Step 1: Update the solution vector by multiplying it by the left-hand factor matrix

$$x_k = l_{kk}^{(k)} \, x_k$$

$$x_i = x_i + l_{ik}^{(k)} \, x_k \qquad i = k+1, \ldots, n$$

Step 2: If $k = n$, then go to Step 4.

Step 3: $k = k + 1$ and go to Step 1.

Right-hand factor matrix by vector multiplications;

Step 4: Set $k = n - 1$.

Step 5: Update the solution vector by multiplying it by the right-hand factor matrix

$$x_k = x_k + \sum_{j=k+1}^{n} r_{kj}^{(k)} \, x_j$$

Step 6: If $k = 1$, then stop.

Step 7: $k = k - 1$ and go to Step 5.

6.2 IMPLEMENTATION OF SPARSE MATRIX TECHNIQUES FOR MICROWAVE NETWORK EQUATIONS WITH WAVE VARIABLES

To exploit an important characteristic of systems of equations (6.1) to (6.3) with an identical zero–nonzero structure of the coefficient matrix when these systems tend to be solved repeatedly many times, it becomes necessary to create a sparse matrix solver that consists of a number of separate segments (subroutines) that are executed sequentially and then repeated in a manner dependent upon the problem solved.

Figure 6.1 presents a structure of a sparse matrix solver for microwave network equations with the connection scattering matrix as a coefficient matrix that uses a looping indexed code technique in which appropriate routines simulate matrix bi-factorization and generate a set of ordering arrays and indexing vectors that can be used to process further matrices of the same zero–nonzero structure. This integer information is used to perform the numeric bi-factorization and finally the numeric solution of circuit equations with the coefficient matrix (connection scattering matrix **W**) of the same zero–nonzero structure. Only arithmetic operations with nonzero elements of **W** are performed [6–8].

The separate routines that create the sparse matrix solver are:

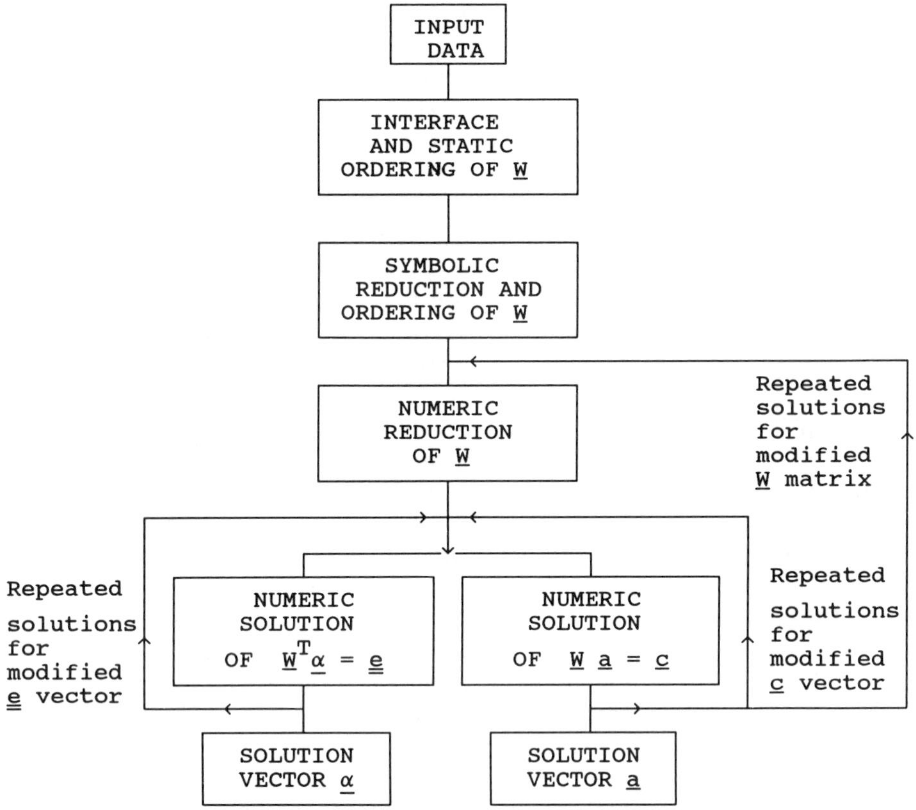

Figure 6.1 Distinct segments of the sparse matrix solver. Different kinds of network analysis are performed by executing only particular segments of the solver.

1. The subroutine for generating the indexing and addressing data and ordering data.
2. The subroutine for the dynamic ordering and simulation of matrix reduction (bi-factorization).
3. The subroutine for numeric reduction.
4. The subroutine for numerically computing the solution vector of the $\mathbf{W}\,\mathbf{a} = \mathbf{c}$ equation.
5. The subroutine for numerically computing the solution vector of the $\mathbf{W}^{\mathrm{T}}\,\alpha = \mathbf{e}_r$ equation.

The first segment of the sparse matrix solver interprets the topological data of the analyzed circuit and generates integer type arrays containing indexing and addressing information describing the zero–nonzero structure of the connection scattering matrix of the microwave circuit to be analyzed. This subroutine interfaces the input data, describing the topology of the circuit to be analyzed with the complicated data structure that is used in the next sparse matrix subroutines. Static ordering that renumbers the equations is also performed in this subroutine. The execution of this subroutine is performed only once for a given circuit with constant topology.

The second routine performs the simulation of the reduction of the coefficient matrix of the system of equations. By the step-by-step simulation of the matrix reduction, the subroutine determines the nonzero structure of factor matrices $\mathbf{L}^{(k)}$ and $\mathbf{R}^{(k)}$, $k = 1, 2, \ldots, k$. The optimal ordering of the matrix is performed before each step of a symbolic reduction of the matrix. The indexing and addressing data contained in the arrays are modified to the form that determines the zero–nonzero structure of the left-hand and right-hand factor matrices. For a given zero–nonzero structure of the coefficient matrix **W**, this subroutine is performed only once.

The third subprogram uses the indexing and addressing information received from the simulation and ordering subprogram and performs the numerical reduction of the actual coefficient matrix. The subprogram performs only the nonzero operations, leading to the computation of nonzero elements of the left-hand $\mathbf{L}^{(k)}$ and right-hand $\mathbf{R}^{(k)}$ factor matrices. This subroutine is invoked each time the numeric values of the **W** matrix entries are changed.

The fourth segment of the solver computes the solution vector. The actual right-hand side vector of the system of circuit equations is stepwise transformed to the solution vector by successive factor matrices by vector multiplications. In the numerical solution routine, only the arithmetic operations with nonzero elements of the **R** and **L** factor matrices are performed. The zero–nonzero structures of the vectors **c**, $\mathbf{e}_r$, and $\mathbf{e}_l$ are not taken into account in the solution vector computation.

The fifth subroutine of the sparse matrix solver computes the solution vector for a system of equations with a transposed connection scattering matrix as a coefficient matrix. The zero–nonzero structures of the right-hand side vector are not taken into account in the solution vector computation.

For a given microwave circuit described by a particular connection scattering matrix **W**, two first segments of the JADMIC sparse matrix solver are executed only once. The repeated analysis of the same circuit is performed by executing only reduction and solution subprograms (frequency domain analysis, change of circuit parameters, sensitivity analysis, and noise analysis) or only the solution subprogram (sensitivity analysis, noise analysis, noise power sensitivity analysis, and change of the location of the generator port in the circuit).

All details of the sparse matrix techniques that very effectively solve microwave network equations with a connection scattering matrix as a coefficient matrix are presented in [8]. Connection scattering matrix ordering strategy and storage scheme of the connection scattering matrix are discussed. Schemes of all the procedures that create the sparse matrix solver are also given.

The sparse matrix solver significantly reduces the amount of memory required to store the matrix **W** and the execution time of the repeated analysis of microwave circuits.

Large systems of equations **W a** = **c** that are solved using conventional full matrix algorithms require storage for $n^2 + n$ complex numbers and execution of the order of n^3 long arithmetic operations. For connection scattering matrices of an order bigger than 10, the amount of computer memory required by the sparse matrix data structure becomes smaller than the amount of memory used for storage of the full matrix. In the case of the sparse matrix solver discussed here, memory requirements as well as execution time vary approximately linearly with the problem size.

References

[1] Forsythe, G. E., and C. M. Moler, *Computer Solution of Linear Algebraic Equations,* Englewood Cliffs, NJ: Prentice-Hall, 1967.

[2] Golub, G. H., and C. F. Van Loan, *Matrix Computation,* The Johns Hopkins University Press, 1983.

[3] Strong, G., *Linear Algebra and Its Applications,* Academic Press, 1980.

[4] Calahan, D. A., *Computer-Aided Network Design,* New York: McGraw-Hill, Inc., 1972.

[5] Vlach, J., and K. Singhal, *Computer Methods for Circuit Analysis and Design,* New York: Van Nostrand Reinhold, 1983.

[6] Zollenkopf, K., "Bi-factorisation—Basic Computational Algorithm and Programming Techniques," *Large Sparse Sets of Linear Equations,* J. K. Reid, Ed., *Proc. Oxford Conference Institute of Mathematics and Its Applications,* April 1970, New York: Academic Press, 1971.

[7] Dobrowolski, J. A., "Algorithms and Storage Scheme in the Sparse Matrix Approach to Computer-Aided Analysis of Microwave Circuits," *Proc. Conf. Computer-Aided Design of Electronic and Microwave Circuits and Systems,* Hull, England, 1977, pp. 122–127.

[8] Dobrowolski, J. A., *Introduction to Computer Methods for Microwave Circuit Analysis and Design,* Norwood, MA: Artech House, 1991.

CHAPTER 7

Microwave Network Optimization Techniques

A very important class of problems in the computer-aided design (CAD) of microwave circuits concerns techniques for adjusting circuit parameters in order to minimize the deviation between the circuit performance achieved at some stage of the design and the desired specifications. For example, the design of a microwave transistor amplifier can have performance functions such as the gain, the bandwidth, the noise figure, and perhaps the input and output impedances. Such techniques are used to refine iteratively an initial design, usually conceived by intuitive methods, until specifications are met. This approach yields results that are otherwise unobtainable and offers particularly attractive economic results in situations involving excluding specifications. The refinement process is an optimization process involving the minimization of a performance criterion that represents a scalar measure of the difference between existing and desired response characteristics of the designed network.

7.1 DEFINITION OF MICROWAVE NETWORK OPTIMIZATION PROBLEM

Optimization, in general, means the process of determining the extreme value (maximum or, more often, minimum) of a mathematical function. The optimization problem is to minimize the scalar function [1–5]

$$U = U(\mathbf{x}) \tag{7.1}$$

subject to the inequality constraints

$$\mathbf{c}(\mathbf{x}) \leq \mathbf{0} \tag{7.2}$$

and equality constraints

$$\mathbf{h}(\mathbf{x}) = \mathbf{0} \tag{7.3}$$

In microwave network design problems, the dependence of the functions $U(\mathbf{x})$, $\mathbf{c}(\mathbf{x})$, and $\mathbf{h}(\mathbf{x})$ on the design parameters $\mathbf{p} = \mathbf{x}$ is implicit through the circuit equations. Thus the objective function evaluations that are required to run an optimization algorithm involve an often expensive simulation of the designed microwave circuit.

The desired performance of the designed microwave network may be expressed by a set of specifications that are functions of the independent variable frequency. Because the response functions can be computed at a finite number of values of frequency, we must consider for such a discrete set of frequencies that satisfying the specifications at these points guarantees satisfying them between the points as well. In general, we can simultaneously consider more than one type of response and can define more than one specification for each of the responses. In practice, the design problem must be defined by lower and upper specifications.

Figure 7.1 presents an example of the lower and upper specifications for the insertion loss characteristic of a band pass filter and the response function that violates these specifications. In this example, f_l and f_u denote, respectively, the lower bound and upper bound of the frequency interval in which the problem is considered.

Figure 7.2 shows an example of response specifications in which there are lower and upper specifications defined for the same interval of the independent variable (that is, frequency).

Another type of specification is a single specification, which is a two-sided specification. Many different design problems may be defined using a single specification. For example, a single specification may be interpreted as a "window" specification having zero width. An example of a single specification for the design problem of an amplifier is shown in Figure 7.3 and describes the requirements on the shape and magnitude of the transducer power gain of the amplifier.

Response functions considered in general purpose programs for microwave network design are:

- Magnitudes of the overall network scattering parameters

$$|S_{11\ \mathrm{net}}|,\ |S_{12\ \mathrm{net}}|,\ |S_{21\ \mathrm{net}}|,\ |S_{22\ \mathrm{net}}|;$$

- Arguments of the overall network scattering parameters

$$\arg\{S_{11\ \mathrm{net}}\},\ \arg\{S_{12\ \mathrm{net}}\},\ \arg\{S_{21\ \mathrm{net}}\},\ \arg\{S_{22\ \mathrm{net}}\};$$

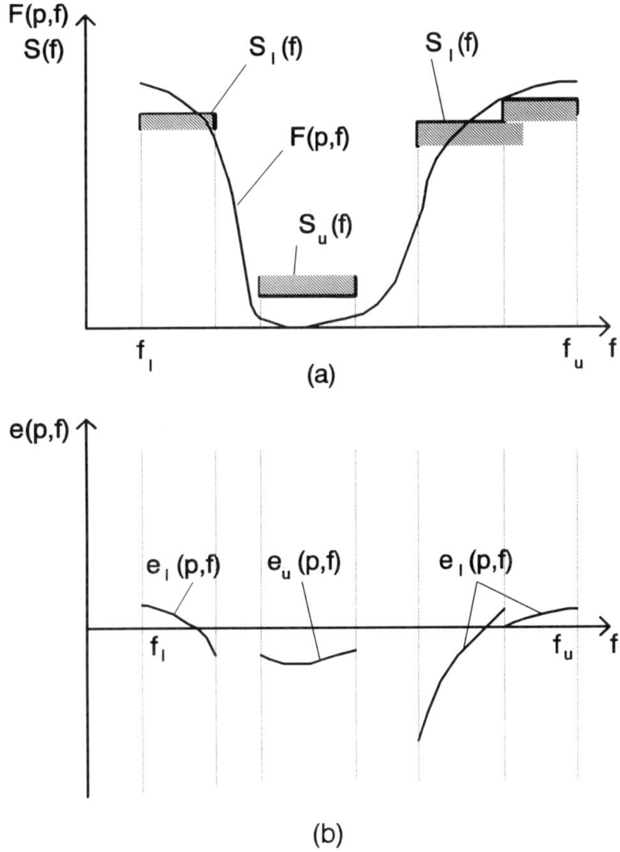

Figure 7.1 (a) Lower $S_l(f)$ and upper $S_u(f)$ specifications imposed on a response function $F(\mathbf{p},f)$ of a band-pass filter. (b) Error function defined as a difference between calculated response and the given specifications.

- Real parts of the overall network scattering parameters

$$\mathrm{Re}\{S_{11\ \mathrm{net}}\}, \mathrm{Re}\{S_{12\ \mathrm{net}}\}, \mathrm{Re}\{S_{21\ \mathrm{net}}\}, \mathrm{Re}\{S_{22\ \mathrm{net}}\};$$

- Imaginary parts of the overall network scattering matrix parameters

$$\mathrm{Im}\{S_{11\ \mathrm{net}}\}, \mathrm{Im}\{S_{12\ \mathrm{net}}\}, \mathrm{Im}\{S_{21\ \mathrm{net}}\}, \mathrm{Im}\{S_{22\ \mathrm{net}}\};$$

- Group delay of the overall network S_{21} transmission coefficient τ_D;
- Noise figure of the overall network F.

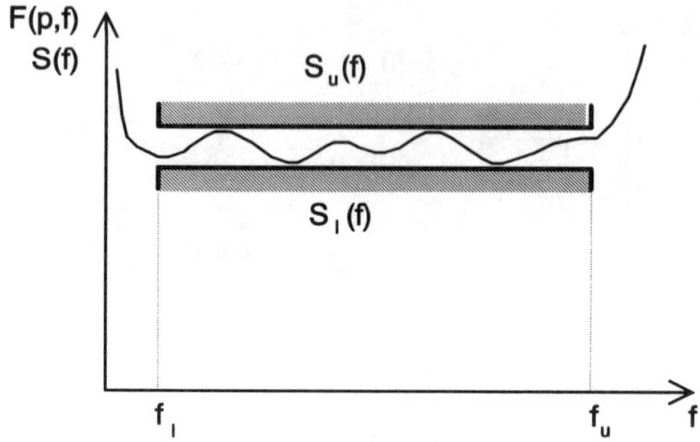

Figure 7.2 Lower $S_l(f)$ and upper $S_u(f)$ specifications for the response function defined for the same interval of frequency.

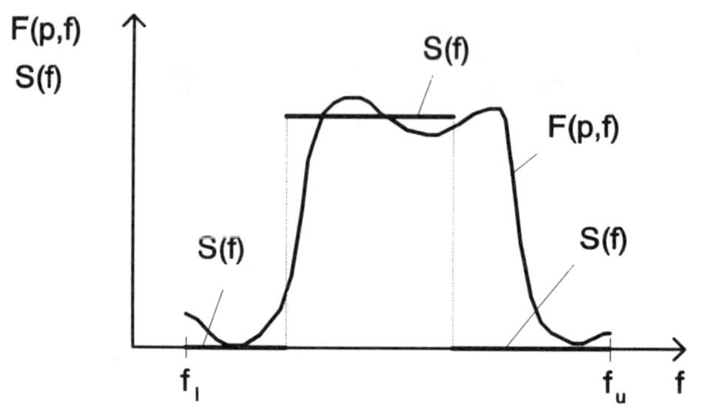

Figure 7.3 The single specifications $s(f)$ and a response function $F(\mathbf{p},f)$ for an amplifier design problem.

Because response functions are calculated at discrete frequencies, it is convenient to define the set of sampled network response functions

$$F_j(\mathbf{p}) \qquad j \in J \tag{7.4}$$

and their corresponding set of sampled specifications

$$S_j \qquad j \in J \tag{7.5}$$

In (7.4) and (7.5) the subscript j refers to quantities (already defined) evaluated at discrete values of the independent variable (that is, discrete frequency points).

Error functions arise from the difference between the actually calculated responses and the given specifications.

In the general case, when we have lower and upper specifications, we define the error functions as [5]

$$e_{uj}(\mathbf{p}) = w_{uj}\,(F_j(\mathbf{p}) - S_{uj}) \qquad j \in J_u \tag{7.6}$$

$$e_{lj}(\mathbf{p}) = w_{lj}\,(F_j(\mathbf{p}) - S_{lj}) \qquad j \in J_l \tag{7.7}$$

where $F_j(\mathbf{p})$ is the response function, $\mathbf{p}$ represents the network parameters, S_{uj} is an upper specification, S_{lj} is a lower specification, w_{uj} is a nonnegative weighting factor for S_{uj}, and w_{lj} is a nonnegative weighting factor for S_{lj}.

Weighting factors w_{uj} and w_{lj} can be thought of as giving the relative importance of $e_{u,lj}(\mathbf{p})$. Simultaneously, each $w_{u,lj}$ must take into account the relative magnitudes or scales associated with $e_{uj,lj}$, which may be defined for different kinds of responses and in different units of measure.

The index sets J_u and J_l defined as

$$J_u = \{j_1, j_2, \ldots, j_k\} \tag{7.8}$$

$$J_l = \{j_{k+1}, j_{k+2}, \ldots, j_m\} \tag{7.9}$$

are not necessarily disjoint.

By redefining the error functions as [5]

$$e_i(\mathbf{p}) = \begin{cases} e_{uj}(\mathbf{p}) & j = j_i \quad i = 1, 2, \ldots, k \\ -e_{lj}(\mathbf{p}) & j = j_i \quad i = k+1, k+2, \ldots, m \end{cases} \tag{7.10}$$

we get a set of uniformly indexed error functions.

The response functions $F(\mathbf{p})$ corresponding to the above defined error functions can only be real-valued functions. Obviously, a positive value of the error function e_i indicates a violation of the corresponding specifications.

If we define

$$M_e(\mathbf{p}) = \max_{i \in I} e_i(\mathbf{p}) \qquad I = \{1, 2, \ldots, m\} \tag{7.11}$$

then the sign of $M_e(\mathbf{p})$ indicates whether all the upper and lower specifications are satisfied or violated; that is, if

$$M_e(\mathbf{p})\begin{cases} >0 & \text{the specifications are violated} \\ =0 & \text{the specifications are exactly met} \\ <0 & \text{the specifications are satisfied with excess} \end{cases}$$

In the case of design problems with single specifications, the definition of error functions is much simpler:

$$e_i = w_i \mid F_i(\mathbf{p}) - S_i \mid \qquad i = 1, 2, \ldots, m \tag{7.12}$$

where w_i is a nonnegative weighting factor.

In (7.12) the response functions and the specifications can be real or complex.

The error functions e_i, $i = 1, 2, \ldots, m$, can be treated as the components of a vector defined as

$$\mathbf{e} = \begin{bmatrix} e_1(\mathbf{p}) \\ e_2(\mathbf{p}) \\ \vdots \\ e_m(\mathbf{p}) \end{bmatrix} \tag{7.13}$$

7.1.1 Generalized l_p Objective Function

In the case when we compute the error functions with respect to the upper and lower specifications, they may have positive as well as negative values. A negative value of $e_i(\mathbf{p})$ means that at this point the specification has been fulfilled with excess. This consideration leads to the most general case, in which some of the e_i for $i \in I$ are nonnegative or all of the e_i for $i \in I$ are negative. If I^+ is a set of i corresponding to nonnegative error functions

$$I^+ = \{ i \mid e_i \geq 0, \ i \in I\} \tag{7.14}$$

then the generalized l_p function has the form [6, 7]

$$U_p(\mathbf{p}) = \begin{cases} U_p^+(\mathbf{p}) & \text{if the set } I^+ \text{ is not empty} \\ U_p^-(\mathbf{p}) & \text{otherwise} \end{cases} \tag{7.15}$$

where

$$U_p^+(\mathbf{p}) = \left[\sum_{i \in I^+} (e_i(\mathbf{p}))^p \right]^{1/p} \tag{7.16}$$

and

$$U_p^-(\mathbf{p}) = -\left[\sum_{i=1}^{m} (-e_i(\mathbf{p}))^{-p}\right]^{-1/p} \tag{7.17}$$

In other words, $U_p^+(\mathbf{p})$ is formed only from the nonnegative error functions and $U_p^-(\mathbf{p})$ is formed from all the error functions when all the e_i are negative, permitting the objective function to be minimized even after all the specifications have been met, allowing the response function to be further improved.

Because, for $e_i > 0$, $i \in I$,

$$\max_{i \in I} \{e_i\} = \lim_{p \to \infty} \left[\sum_{i \in I} (e_i)^p\right]^{1/p} \tag{7.18}$$

and

$$\min\{e_i\} = \lim_{p \to \infty} \left[\sum_{i \in I} (e_i)^{-p}\right]^{-1/p} \tag{7.19}$$

it is comprehensible that by minimizing the generalized l_p function given by (7.15) to (7.17) with very large values of p, we obtain results that are very close to the minimax optimum.

The generalized l_p function without some modifications would not be applicable in practice due to ill-conditioning that results from the numerical evaluation of $(\mp\, e_i)^{\mp p}$ for very large values of p. This ill-conditioning problem may be solved simply by dividing all the error functions e_i by a value equal to the largest magnitude error function M_e. Using this idea we define the objective function in the following manner [6, 7]:

$$U(\mathbf{p}) = \begin{cases} M_e \left[\sum_{i \in K} \left(\dfrac{e_i(\mathbf{p})}{M_e}\right)^q\right]^{1/q} & \text{for } M_e \neq 0 \\ 0 & \text{for } M_e = 0 \end{cases} \tag{7.20}$$

where

$$q = \frac{M_e}{|M_e|}\, p \begin{cases} 1 < p < \infty & \text{for } M_e > 0 \\ 1 \le p < \infty & \text{for } M_e < 0 \end{cases} \tag{7.21}$$

$$K = \begin{cases} I^+ & \text{if } M_e > 0 \\ I & \text{if } M_e < 0 \end{cases} \tag{7.22}$$

In (7.19) to (7.21)

$$M_e = \begin{cases} \lim\limits_{p \to \infty} U_p^+ (\mathbf{p}) & \text{if the set } I^+ \text{ is not empty} \\ \lim\limits_{p \to \infty} U_p^- (\mathbf{p}) & \text{otherwise} \end{cases} \tag{7.23}$$

The gradient vector of the objective function is given by

$$\underline{\nabla}\, U(\mathbf{p}) = \left[\sum_{i \in K} \left(\frac{e_i(\mathbf{p})}{M_e} \right)^q \right]^{\frac{1}{q} - 1} \cdot \sum_{i \in K} \left(\frac{e_i(\mathbf{p})}{M_e} \right)^{q-1} \underline{\nabla} e_i(\mathbf{p}) \qquad \text{for } M_e \neq 0 \tag{7.24}$$

It is observed from (7.20) and (7.24) that if $e_i(\mathbf{p})$ for $i \in I$ is a set of continuous functions with continuous first partial derivatives, then the objective function U and its first partial derivative are also continuous. The discontinuities in the derivative of the objective function of (7.20) occur only at a point $\mathbf{p}$ where $M_e(\mathbf{p}) = 0$, and two or more maxima are equal. The minimization of the objective function defined by (7.20) to (7.23) can be performed using gradient optimization methods that are more efficient than direct search methods.

The objective function $U(\mathbf{p})$ is an l_p norm of the error functions $e_i(\mathbf{p})$ given by (7.12). The l_p norm of $\mathbf{e}(\mathbf{p})$ is defined as [8]

$$\|\mathbf{e}\|_p = \left\{ \sum_{i=1}^{m} |e_i(\mathbf{p})|^p \right\}^{1/p} \qquad \begin{matrix} i \in I \\ 1 \leq p \leq \infty \end{matrix} \tag{7.25}$$

where the l_p norm is a scalar measure of the deviations between the desired and the actual responses of the designed circuit.

The choice of the value of the parameter p in the l_p norm has an essential significance. By taking a small value for p we in effect emphasize those error functions e_i that have smaller values. When the parameter p has a large value, these error functions e_i that have larger values decide on the value of the l_p norm. Three cases of the l_p norm—$p = 1$, $p = 2$, and $p = \infty$—are the most well known and widely used in microwave network optimization.

The l_1 Norm

The l_1 norm, defined as

$$\|\mathbf{e}\|_1 = \sum_{i=1}^{m} |e_i(\mathbf{p})| \tag{7.26}$$

emphasizes the importance of these error functions e_i that are close to zero. The l_1 norm is used as the objective function in the approximation problem defined as

$$\underset{\mathbf{p}}{\text{minimize}} \|\mathbf{e}(\mathbf{p})\|_1 = \sum_{i=1}^{m} |e_i(\mathbf{p})| \tag{7.27}$$

The l_1 norm, as compared to the other norms l_p with $p > 1$, has the distinctive property that some components of **e** with large values are ignored; that is, at the solution of (7.15) there may be a few e_is that are much larger than the others. The solution of (7.15) is usually a point where one or more of the e_is are equal to zero while some large e_is are completely ignored. This property is very important in problems of approximating data that might contain some points weighted with large errors. It means that the l_1 norm is preferable in the problems of approximating a measured response by a response of a device, a network, or a system formulated as an optimization problem with respect to the equivalent circuit parameters of a proposed model [9, 10]. Because of the fact that $\|\mathbf{e}\|_1$ is not differentiable in the ordinary sense, their minimization requires sophisticated algorithms [11–15].

The l_2 Norm (p = 2)

The l_2 norm (p = 2), defined as

$$\|\mathbf{e}\|_2 = \left[\sum_{i=1}^{m} |e_i(\mathbf{p})|^2\right]^{1/2} \tag{7.28}$$

is perhaps the most well-known and widely used norm; it is called the least-squares norm. The l_2 norm that is used as the objective function has various applications in design problems formulated as optimization problems [16–18]. A very important property of the l_2 norm is its differentiability; that is, its gradient vector can be obtained very easily from the partial derivatives of **e**.

The l_∞ Norm ($p = \infty$)

By letting in (7.25) p = ∞ we have $\|\mathbf{e}\|_\infty$ which is called Chebyshev or uniform norm. Because

$$\lim_{p\to\infty}\left\{\sum_{i=1}^{m}|e_i|^p\right\}^{1/p} = \max_i |e_i| \qquad i \in I \tag{7.29}$$

the uniform norm

$$\|\mathbf{e}\|_\infty = \max_i |e_i| \qquad i \in I \tag{7.30}$$

takes a value of the error function e_i with the largest value (the worst case) and the other errors are in effect ignored.

Minimax optimization defined as

$$\min_{\mathbf{p}} \max_i |e_i(\mathbf{p})| \tag{7.31}$$

typically leads to the optimal solution $\mathbf{p}^*$ for which values of some error functions are the same

$$e_{i_1}(\mathbf{p}^*) = e_{i_2}(\mathbf{p}^*) = \cdots = e_{i_k}(\mathbf{p}^*) \tag{7.32}$$

with the other values of $e_i(\mathbf{p}^*)$ less than this value. This equality at the final answer is equivalent to the equal-ripple error characteristic of the circuit function. This type of solution is often required in microwave filter design problems.

7.1.2 Constraints on Optimized Network Parameters

Unconstrained optimization often produces unacceptable parameter values that have no physical or practical meaning.

The subject of constraints deals with the explicit or implicit relationships among optimization parameters **p** of a designed network. The most elementary constraints are bounds on network parameters; these may prevent negative parameter values or contain parameter values within certain ranges.

The most natural constraints are upper and/or lower bounds, which can be written in the form

$$p_{li} \leq p_i \leq p_{ui} \tag{7.33}$$

where p_{li} and p_{ui} are the lower and the upper bounds for the *i*th parameter of the network. These bounds usually come from technological, topological, and electrical considerations imposing restrictions on different network parameters.

There are many methods for constrained optimization. Bounds on optimization parameters given by (7.33) are best incorporated by transforming the variables. For example, allowing the optimization procedure to adjust the optimized parameter x in $p = x^2$ while computing the objective function $U = U(x)$ will maintain p as positive.

A means for maintaining parameter values within lower and upper bounds is in the transformations

$$p_i = A + B\frac{x_i}{1 + |x_i|^2} \tag{7.34}$$

where

$$A = \frac{p_{li} + p_{li}}{2} \tag{7.35}$$

and

$$B = \frac{p_{ui} - p_{li}}{2} \tag{7.36}$$

for all n network parameters.

It is also possible to use a transformation function of the form

$$p_i = A + B \sin x_i \tag{7.37}$$

Variables are transformed outside the optimization subroutine. The user inputs the value of p_i that satisfies the constraints given by (7.33); the optimization subroutine works with the variable x_i, which is the transformed value of p_i. The objective function routine, also being outside the optimization subroutine, converts the variable x_i back to p_i before making its calculations. Therefore, even though in the optimization subroutine its variables can take values outside their bounds, there will be no decrease in the objective function subroutine and the optimization subroutine will move its variables to the acceptable region defined by parameter constraints.

Example 7.1

Figure 7.4 presents a low-pass filter structure as it might appear if designed on a microstrip line. The filter consists of a cascade of connections of microstrip lines of

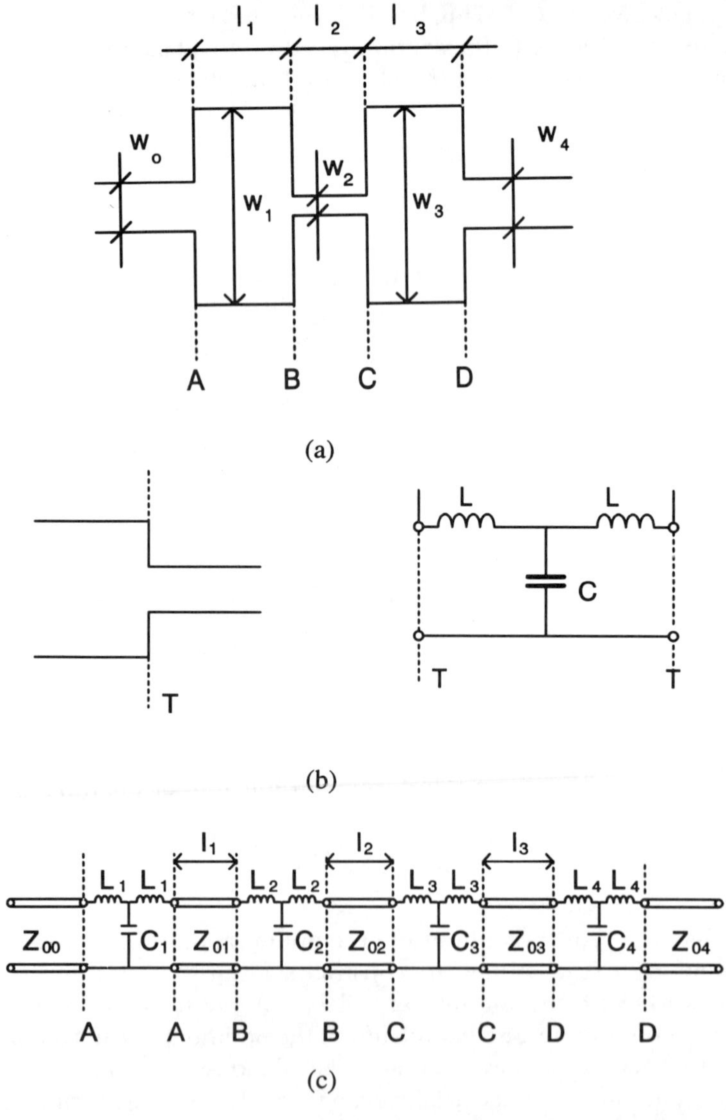

Figure 7.4 Microstrip low-pass filter and its simulation model: (a) the pattern of the strip-line structure of the filter, (b) lumped equivalent circuit of an abrupt step in width of the microstrip line, and (c) equivalent circuit of the low-pass microstrip filter.

low characteristic impedance (the broad strips) and lines of high impedance (the narrow strips). As the low-level model parameters we can define eleven physical parameters of the microstrip filter structure as strip widths of the succeeding microstrip line sections w_0, w_1, w_2, w_3, w_4; their lengths l_1, l_2, l_3; the dielectric constant of a substrate ε_r; the substrate thickness b; and the strip metallization thickness t. The vector $\mathbf{p}^L$ that contains these parameters is

$$\mathbf{p}^L = [w_0, w_1, w_2, w_3, w_4, l_1, l_2, l_3, \varepsilon_r, b, t]^{\mathrm{T}}$$

Electrically, the discussed microstrip low-pass filter structure can be thought of as a cascade of transmission line segments each with different characteristic impedance, phase shift, line loss, and a guide wavelength. Figure 7.4(c) depicts the equivalent circuit of the discussed low-pass filter structure. It takes into account effects related to the discontinuities associated with the abrupt change of strip width. It is a higher level model and the vector $\mathbf{p}^H$ that contains its sixteen parameters is

$$\mathbf{p}^H = [Z_{o0}, Z_{o1}, Z_{o2}, Z_{o3}, Z_{o4}, l_1, l_2, l_3, L_1, L_2, L_3, C_1, C_2, C_3, C_4]^{\mathrm{T}}$$

The relationships between the high-level parameters and the physical, low-level parameters of the system are available from the theory of the microstrip line and its discontinuities.

In practical design we can assume the symmetry of the filter structure. In that way the number of high-level design parameters may be reduced to nine. Because $Z_{o0} = Z_{o4}$, $Z_{o1} = Z_{o3}$, $l_1 = l_3$, $L_1 = L_4$, $L_2 = L_3$, $C_1 = C_4$, $C_2 = C_3$, we have

$$\mathbf{p} = [Z_{o0}, Z_{o1}, Z_{o2}, l_1, l_2, L_1, L_2, C_1, C_2]^{\mathrm{T}}$$

The responses $\mathbf{F}$ of the designed low-pass filter structure can be the frequency characteristics of the transmission loss $L(\omega) = 20 \log(|S_{21}(j\omega)|^{-1})$ and of the group delay $\tau = -\partial/\partial\omega(\arg\{S_{21}\})$.

Example 7.2

Let us consider a single-stage transistor amplifier design. Figure 7.5 illustrates an equivalent circuit of the amplifier. Two simple matching networks with shunt short-circuited stubs are used at the input and output of the transistor. Characteristic impedances and lengths of transmission lines creating matching circuits are design parameters of the discussed design problem

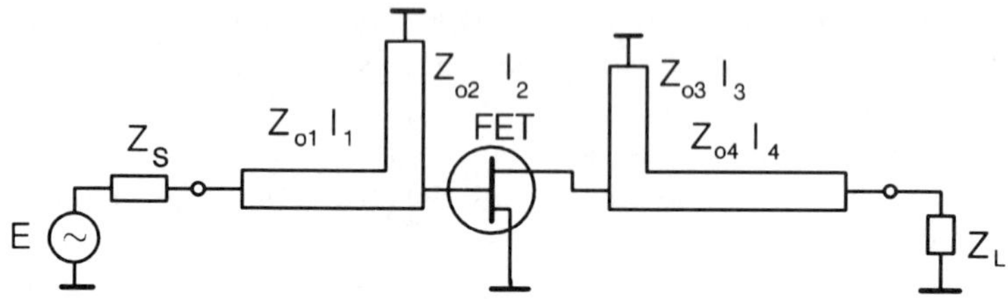

Figure 7.5 Equivalent network of a single-stage transistor amplifier. Characteristic impedances Z_{O1}, Z_{O2}, Z_{O3}, Z_{O4} and lengths l_1, l_2, l_3, l_4 of transmission lines creating input and output matching circuits create a set of design parameters for the transistor amplifier design problem.

$$\mathbf{p} = [Z_{o1}, l_1, Z_{o2}, l_2, Z_{o3}, l_3, Z_{o4}, l_4]^T$$

The scattering matrix parameters related to the input and output ports of the amplifier characterize the whole circuit and are functions of the design parameters. When Z_S and Z_L equal the reference impedances, respectively, of the input and of the output ports of the amplifier, $|S_{21}|^2$ represents transducer power gain of the amplifier, S_{11} and S_{22} are the input and output reflection coefficients, and $|S_{12}|^2$ is the backward transducer power gain of the amplifier.

These resultant parameters are utilized to create error functions and, finally, the objective function. We define the error functions:

$$e_i(\mathbf{p}) = |S_{21}(\mathbf{p},f_i)|^2 - |S_{21}^s(\mathbf{f}_i)|^2$$
$$e_i(\mathbf{p}) = |S_{11}(\mathbf{p},f_i)|$$
$$e_i(\mathbf{p}) = |S_{22}(\mathbf{p},f_i)|$$

where $i \in I \triangleq \{1, 2, \ldots, m\}$ is a set of integers corresponding to a number of discrete frequency points f_i in a frequency band of interest and $|S_{21}^s(f_i)|^2$ is the specified objective for the power gain.

The objective function can be modified to stress various aspects of the amplifier design, gain flatness, maximum power transfer, minimum reflection on the input and output of the amplifier, minimum noise figure, and various combinations of these features.

Using the objective function of the form

$$U(\mathbf{p}) = \sum_{i=1}^{m} \{w_{21i} ||S_{21}(\mathbf{p},f_i)|^2 - |S_{21}^s(f_i)|^2|^p + w_{11i}|s_{11}(\mathbf{p},f_i)|^p + w_{22i}|S_{22}(\mathbf{p},f_i)|^p\}$$

and minimizing it with respect to circuit parameters **p** correspond to a compromise between power gain and good match at the input and output of the amplifier. In $U(\mathbf{p})$ the w_{21i}, w_{11i}, and w_{22i} are positive weighing factors.

When designing the amplifier with optimum noise performance, it is possible to utilize a two-step procedure. In the first step we design the input matching network to provide the best noise performance of the transistor. The appropriate objective function has the form

$$U(\mathbf{p}) = \sum_{i=1}^{m} w_i (F(\mathbf{p}, f_i))^p$$

where $F(\mathbf{p}, f_i)$ is a noise figure of the amplifier and $\mathbf{p}$ is a vector of design parameters given as

$$\mathbf{p} = [Z_{o1}, l_1\ Z_{o2}, l_2]$$

In the second step the output matching network is designed to provide maximum constant gain in the frequency band of interest.

Example 7.3

In general-purpose computer programs the objective functions usually have the form

$$U(\mathbf{p}) = \sum_{j=1}^{k} \left\{ \sum_{i=1}^{l_j} \sum_{m=1}^{6} [w_{ij}^{(m)} (F_{ij}^{(m)}(\mathbf{p}) - S_{ij}^{(m)})]^p \right\}$$

where j denotes the optimization frequency band 1 to k, i the frequency points 1 to l_j in a frequency band j, m the circuit function being optimized, $w_{ij}^{(m)}$ the weight for the circuit function m at a frequency point point i in band j, $F_{ij}^{(m)}(\mathbf{p})$ the value of the circuit function m at a frequency point i in band j, $S_{ij}^{(m)}$ the desired value of the circuit function m at a frequency point i in band j, and $\mathbf{p}$ the circuit parameters.

The six circuit functions are usually four overall scattering parameters referred to the input and output ports of a circuit, the noise figure, and group delay. Because scattering parameters of a circuit are complex, their real and imaginary parts have to be controlled separately. To do this components of the objective function $U(\mathbf{p})$ corresponding to overall scattering parameters should be computed as

$$|w_{ij}^{(m)} \mathrm{Re}\{F_{ij}^{(m)}(\mathbf{p}) - S_{ij}^{(m)}\}|^p + |w_{ij}^{(m)} \mathrm{Im}\{F_{ij}^{(m)}(\mathbf{p}) - S_{ij}^{(m)}\}|^p$$

Instead of controlling real and imaginary parts of complex scattering parameters, it is possible to control their magnitudes and phases. In such a case, components of the sum forming the objective function will have the form

$$|w_{ij}^{(m)}\,(|F_{ij}^{(m)}\,(\mathbf{p})| - |S_{ij}^{(m)}|)|^{\,p} + |w_{ij}^{(m)}\,(\arg\{F_{ij}^{(m)}\,(\mathbf{p})\} - \arg\{S_{ij}^{(m)}\})|^{p}$$

The choice depends on the problem to be solved. If the sign of the real and/or imaginary part of the circuit function is essential, the first approach should be used.

The values of scattering parameters differ in magnitude. For passive circuits they range from zero to unity, and for active circuits they are allowed to be bigger than one. In general optimization problems, there may be a wide variance in values of different circuit functions or in values of one function for different frequency bands. So that all circuit functions should contribute adequately to the objective function, weights $w_{ij}^{(m)}$, which may be different for each circuit function and each frequency band, are allowed.

References

[1] Gill, P. E., W. Murray, and M. H. Wright, *Practical Optimization,* San Diego, CA: Academic Press, 1981.

[2] Gill, P. E., W. Murray, and M. H. Wright, *Numerical Linear Algebra and Optimization,* Redwood City, CA: Addison-Wesley Publishing Co., 1991.

[3] Bandler, J. W., "Optimization Methods for Computer Aided Design," *IEEE Trans. Microwave Theory Tech.,* Vol. MTT-17, August 1969, pp. 533–552.

[4] Bandler, J. W., "Computer-Aided Circuit Optimization," *Modern Filter Theory and Design,* G. C. Temes and S. K. Mitra, Eds., New York: Wiley, 1973, pp. 211–271.

[5] Bandler, J. W., and S. H. Chen, "Circuit Optimization: the State of the Art," *IEEE Trans. Microwave Theory Tech.,* Vol. MTT-36, February 1988, pp. 424–443.

[6] Bandler, J. W., and C. Charalambous, "Theory of Generalized Least pth Approximation," *IEEE Trans. Circuit Theory,* Vol. CT-19, 1974, pp. 607–619.

[7] Charalambous, C., "Nonlinear Least pth Optimization and Nonlinear Programing," *Math. Prog.,* Vol. 12, 1977, pp. 195–225.

[8] Temes, G. C., and D. Y. F. Zai, "Least *p*th Approximation," *IEEE Trans. Circuit Theory,* Vol. CT-16, 1969, pp. 235–237.

[9] Bandler, J. W., W. Kellermann, and K. Madsen, "A Nonlinear l_1 Optimization Algorithm for Design, Modeling and Diagnosis of Networks," *IEEE Trans. Circuits Syst.,* Vol. CAS-34, 1987, pp. 174–181.

[10] Bandler, J. W., S. H. Chen, and S. Daijavad, "Microwave Device Modeling Using Efficient l_1 Optimization: a Novel Approach," *IEEE Trans. Microwave Theory Tech.,* Vol. MTT-34, 1986, pp. 1282–1293.

[11] Hald, J., and K. Madsen, "Combined LP and Quasi-Newton Methods for Nonlinear l_1 Optimization," *SIAM J. Numer. Anal.,* Vol 22, 1985, pp. 68–80.

[12] Hald, J., *A 2-Stage Algorithm for Nonlinear l_1 Optimization,* Report No. NI-81–01, Inst. for Numerical Analysis, Tech. Univ. of Denmark, Lyngby, 1981.

[13] Osborne, M. R., and G. A. Watson, "On an Algorithm for Discrete Nonlinear l_1 Approximation," *Computer J.,* Vol. 14, 1971, pp. 184–188.

[14] Murray, W., and M. L. Overton, "A Projected Lagrangian Algorithm for Nonlinear l_1 Optimization," *SIAM J. Sci. Stat. Comput.,* Vol. 2, 1981, pp. 207–224.
[15] El-Attar, R. A., M. Vidyasagar, and S. R. K. Dutta, "An Algorithm for l_1-Norm Minimization with Application to Nonlinear l_1-Approximation," *SIAM J. Numer. Anal.,* Vol. 16, 1979, pp. 70–86.
[16] Perlman, B. S., and V. G. Gelnovatch, "Computer Aided Design, Simulation and Optimization," *Advances in Microwaves,* L. Young, Ed., New York: Academic Press, 1974.
[17] Hald, J., *MMLA1Q, a FORTRAN Subroutine for Linearly Constrained Minimax Optimization,* Report No. NI-81–01, Inst. for Numerical Analysis, Tech. Univ. of Denmark, Lygby, Denmark, 1981.
[18] Bartels, R. H., A. R. Conn, and J. W. Sinclair, "Minimization Techniques for Piecewise Differentiable Functions: the l_1 Solution to an Overdetermined Linear System," *SIAM J. Numer. Anal.,* Vol. 15, No. 2, 1978, pp. 224–241.

CHAPTER 8

Time Domain Microwave Network Analysis Using Waves and Frequency Domain Data

High-speed digital applications of microwave networks lead to the need for a transient time domain simulator. The analysis of interconnect structures has become an important requirement in the design of high-speed and analog networks [1, 2]. Linear microwave networks can be characterized accurately in the frequency domain. Standard linear network modeling techniques offer an efficient method of analysis. It is well known that specific time domain responses of a linear network can be derived theoretically if its frequency response characteristics are fully known. In the case of a microwave network, these characteristics (data) are in the form of frequency-dependent scattering parameters. Fourier series transformations are commonly used for such determinations.

8.1 FOURIER SERIES AND FOURIER TRANSFORM—MATHEMATICAL BASIS

It is well known that signals can be represented either in the frequency domain or time domain. Specific time domain responses of a linear network can be predicted theoretically if its frequency characteristics are fully known.

The relationship between time and frequency forms of mathematical representation of signals is given by the Fourier series for periodic time functions and by the Fourier transformation for nonperiodic functions.

The Fourier series involves an infinite number of terms. The equation that defines the Fourier series for a periodic time function of period T is

$$f(t) = A^{r}(0) + \sum_{n=1}^{\infty} (A^{r}(n)\cos(n\,\omega_0 t) + B^{r}(n)\sin(n\,\omega_0 t)) \tag{8.1}$$

where

$$A^{r}(0) = \frac{1}{T}\int_{-T/2}^{+T/2} f(t)\,dt \tag{8.2}$$

$$A^{r}(n) = \frac{2}{T}\int_{-T/2}^{+T/2} f(t)\cos(n\,\omega_0 t)\,dt \tag{8.3}$$

$$B^{r}(n) = \frac{2}{T}\int_{-T/2}^{+T/2} f(t)\sin(n\,\omega_0 t)\,dt \tag{8.4}$$

In (8.1), a periodic time function $f(t)$ is represented as the sum of an infinite series of sines and cosines of harmonically related frequencies. In (8.2) to (8.4) $\omega_0 = 2\,\Pi/T$ is the fundamental radial frequency.

The coefficients $A^{r}(0)$, $A^{r}(n)$, and $B^{r}(n)$ are the amplitudes of the harmonic components that constitute the original function $f(t)$. They give the frequency spectrum of the function $f(t)$.

Equation (8.1) may also be written in exponential form as

$$f(t) = \sum_{-\infty}^{+\infty} A^{d}(n)\, e^{j\,n\,\omega_0 t} \tag{8.5}$$

where

$$A^{d}(n) = \frac{1}{T}\int_{-T/2}^{+T/2} f(t)\, e^{-j\,n\,\omega_0 t}\,dt \tag{8.6}$$

This can be done by expressing $\sin(x)$ and $\cos(x)$ in (8.1) as Euler relations

$$\sin(x) = \frac{e^{jx} - e^{-jx}}{2j} \quad \text{and} \quad \cos(x) = \frac{e^{jx} + e^{-jx}}{2} \tag{8.7}$$

In the exponential form of the Fourier series, positive as well as negative frequencies are allowed. The coefficients $A^{d}(n)$ in (8.5) are not amplitudes of sine and cosine functions; they are coefficients of exponential functions. The relationships between real amplitudes $A^{r}(0)$, $A^{r}(n)$, and $B^{r}(n)$ and complex coefficients $A^{d}(n)$ are given as

$$A^{\mathrm{r}}(0) = A^{\mathrm{d}}(0) \tag{8.8}$$

$$A^{\mathrm{r}}(n) = \frac{A^{\mathrm{d}}(n) + A^{\mathrm{d}}(-n)}{2} \tag{8.9}$$

$$B^{\mathrm{r}}(n) = \frac{A^{\mathrm{d}}(-n) - A^{\mathrm{d}}(n)}{2} \tag{8.10}$$

Real coefficients $A^{\mathrm{r}}(n)$, $n = 0, 1, 2, \ldots$, and $B^{\mathrm{r}}(n)$, $n = 1,2, \ldots$, are called real spectrum, while complex coefficients $A^{\mathrm{d}}(n)$, $n = 1, 2, \ldots$, are double-sided spectrum of periodic signal $f(t)$.

In the case of nonperiodic time functions $f(t)$ the period T extends to ∞. In the limit, for $T \to \infty$, from (8.5) and (8.6) we obtain the Fourier transform (FT)

$$F(j\omega) = \int_{-\infty}^{+\infty} f(t)\, e^{-j\omega t}\, dt \tag{8.11}$$

and the inverse Fourier transform (IFT)

$$f(t) = \frac{1}{2\Pi} \int_{-\infty}^{+\infty} F(j\omega)\, e^{j\omega t}\, d\omega \tag{8.12}$$

In this case the frequency spectrum $F(j\omega)$ is a continuous function considered for positive as well as negative frequencies.

The Fourier series and the Fourier transform are linear transformations. They can be applied to linear functions. In a linear network there is no interaction between different applied frequencies, so we can linearly superimpose the effects of many simultaneous inputs of different frequencies.

The real part of the spectrum $F(j\omega) = R(\omega) + j\, X(\omega)$ is an even function of frequency

$$R(\omega) = R(-\omega) \tag{8.13}$$

while imaginary part of $F(j\omega)$ is an odd function of the frequency

$$X(\omega) = -X(\omega) \tag{8.14}$$

From (8.11) we also have

$$R(\omega) = \int_{-\infty}^{+\infty} f(t) \cos(\omega t)\, dt \tag{8.15}$$

and

$$X(\omega) = \int_{-\infty}^{+\infty} f(t) \sin(\omega t) \, dt \tag{8.16}$$

For a two-port presented in Figure 8.1(a), the relationship between the output signal $f_{out}(t)$ and the input signal $f_{in}(t)$ is given by the convolution integral of the input signal $f_{in}(t)$ and the impulse response $h(t)$ of the network [3]

$$f_{out}(t) = f_{in}(t) * h(t) = \int_{-\infty}^{+\infty} f_{in}(t) \, h(t-\tau) \, d\tau \tag{8.17}$$

The function *h(t)* is the response of the network excited at the input by a unit impulse function, δ(t). If the function *h(t)* is known, (8.18) allows calculating the time domain output function for any given input excitation.

In the case of a linear and time invariant network, applying the Fourier

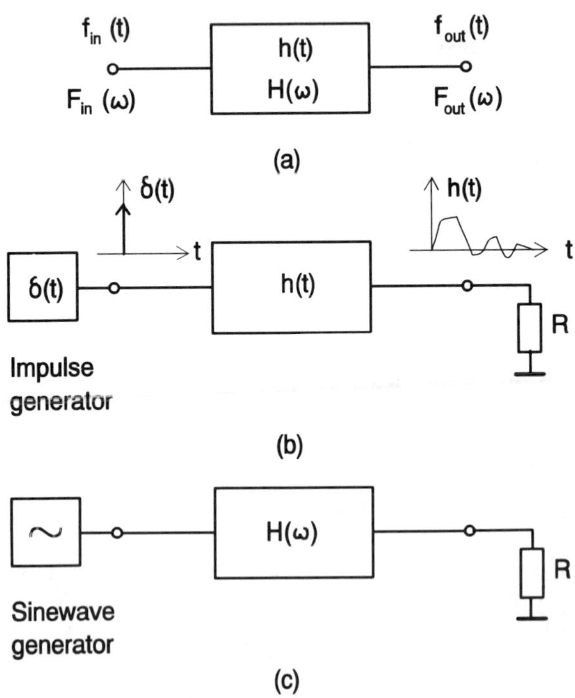

Figure 8.1 (a) Relations between input and output signals of a two-port in time and frequency domain, (b) impulse excitation of a two-port, and (c) sinusoidal excitation of a two-port.

transform to (8.17) leads to the relation between the input function $F_{in}(\omega)$ and the output function $F_{out}(\omega)$ in the frequency domain

$$F_{out}(\omega) = H(\omega)\, F_{in}(\omega) \tag{8.18}$$

$F_{out}(\omega)$ and $F_{in}(\omega)$ in (8.18) are the Fourier transforms of $f_{in}(t)$ and $f_{out}(t)$, respectively, and the network function $H(\omega)$ is the Fourier transform of the impulse response $h(t)$.

8.2 TIME DOMAIN NETWORK ANALYSIS USING WAVES

The solution of

$$\mathbf{W\, a} = \mathbf{c} \tag{8.19}$$

where

$$\mathbf{W} = \mathbf{\Gamma} - \mathbf{S} \tag{8.20}$$

is a vector of ingoing waves **a** at all network ports. In (8.19) **c** is a vector of network's excitations.

A vector of outgoing waves **b** at all network ports can be calculated from the equation

$$\mathbf{b} = \mathbf{\Gamma\, a} \tag{8.21}$$

Port terminal voltages and currents are respectively calculated from the equations

$$\mathbf{U} = \sqrt{Z_R}\,(\mathbf{a} + \mathbf{b}) \tag{8.22}$$

$$\mathbf{I} = \frac{\mathbf{a} - \mathbf{b}}{\sqrt{Z_R}} \tag{8.23}$$

Until this point in our discussion, all the relations presented are considered in the frequency domain.

Assuming that the frequency-dependent scattering parameters of a network are known, the time domain response of the network can be found using the following procedure.

First, assume the network's excitation $\mathbf{c}(t)$ is subject to a discrete fast Fourier transform (FFT) [4]. An N-point fast Fourier transform produces a discrete frequency spectrum $\mathbf{c}(\omega_n)$, $n = 1, 2, \ldots, N$, of the excitation signal $\mathbf{c}(t)$.

The discrete frequency spectrum $\mathbf{c}(\omega_k)$ is computed at discrete frequencies

$$f_n = \frac{n}{2\,N\Delta} \qquad n = 1, 2, \ldots, N \tag{8.24}$$

where Δ denotes the time interval between consecutive samples of $\mathbf{c}(t)$. The extreme value of n in (8.24) corresponds exactly to the Nyquist critical frequency $f_c = 1/(2\Delta)$ [4].

If the spectrum of $\mathbf{c}(t)$ is strictly limited to N harmonics, then according to sampling theory $2N$ samples is sufficient. Practical signals usually have spectrums that are not strictly limited, and the number of samples of $\mathbf{c}(t)$ must be substantially larger to avoid the so-called aliasing errors in the analysis. For example, when the number of harmonics under consideration is 10, the number of samples may equal 128.

Second, using the standard procedure of frequency domain network analysis, discrete frequency spectrums of the ingoing and outgoing waves are computed from the equations

$$\mathbf{a}(\omega_n) = \mathbf{W}^{-1}\,\mathbf{c}(\omega_n) \tag{8.25}$$

$$\mathbf{b}(\omega_n) = \mathbf{\Gamma}\,\mathbf{a}(\omega_n) \tag{8.26}$$

Finally, the time domain response of the network is found by performing the inverse fast Fourier transform (IFFT) of the sum and/or difference of $\mathbf{a}(\omega_n)$ and $\mathbf{b}(\omega_n)$. According to (8.22) and (8.23) the sum and the difference of $\mathbf{a}$ and $\mathbf{b}$ represent port terminal voltages and currents, respectively. Voltage and current network responses in the time domain are computed as

$$\mathbf{V}(t) = \sqrt{Z_R}\ \mathrm{IFFT}\ \{\mathbf{a}(\omega_n) + \mathbf{b}(\omega_n)\} \tag{8.27}$$

$$\mathbf{I}(t) = \frac{1}{\sqrt{Z_R}}\ \mathrm{IFFT}\ \{\mathbf{a}(\omega_n) - \mathbf{b}(\omega_n)\} \tag{8.28}$$

References

[1] Cooke, B. J., J. L. Prince, and A. C. Calgellaris, "S-Parameter Analysis of Multiconductor, Integrated Circuit Interconnect Systems," *IEEE Trans. Microwave Theory Tech.*, Vol. 11, No. 3, March 1992, pp. 353–360.

[2] Dhaene, T., L. Martens, and D. De Zutter, "Transient Simulation of Arbitrary Nonuniform Inter-

connection Structures Characterized by Scattering Parameters," *IEEE Trans. Microwave Theory Tech.,* Vol. 11, No. 11, November 1992, pp. 928–937.

[3] Champeney, D. C., *Fourier Transforms and Their Physical Applications,* New York: Academic Press, 1973.

[4] Elliott, D. F., and K. R. Rao, *Fast Transforms: Algorithms, Applications,* New York: Academic Press, 1982.

CHAPTER 9

Nonlinear Microwave Network Analysis Using Waves

The computer-aided microwave analysis methods presented in Chapters 3, 4, and 5 are applicable to microwave networks driven by small signals. In these cases, microwave semiconductor devices such as Schottky diodes and BJT or FET transistors are linear and, therefore, a whole network can be considered as a linear network.

For the analysis and design of such networks as high-power transistor amplifiers, oscillators, frequency multipliers, and mixers, it is necessary to apply a different approach. The large signal operation of networks causes semiconductor devices to behave nonlinearly, which completely changes the CAD methods that are used for network analysis.

The nonlinear analysis of a microwave network corresponds to the simulation of its state when the network is driven by a source of a periodic large signal of an angular frequency ω_o and by DC sources biasing semiconductor devices. In microwave nonlinear networks only a periodic steady-state response of the frequency ω_o is of practical interest. The aim of the nonlinear analysis is to determine the large signal properties of the network, such as a power delivered to the load, power-aided efficiency, and an input port reflection coefficient.

There are many ways to find a steady-state response of a nonlinear network excited at a given frequency ω_o [1–6], and they are usually divided into two groups. In the first group, called *transient methods,* the network behavior is simulated after the periodic excitation is turned on and until the steady state is reached. The network equations are formulated in the time domain and then integrated in time until the steady state is reached [1]. Transient methods simulate real behavior of a network. They may be used to investigate dynamic properties of nonlinear networks, such as response to a step change of an amplitude or phase of a driving signal.

The second group of methods, called *balanced methods,* find the periodic solution of a network directly, by discretizing, in the frequency or in the time domain, the periodic continuous signals involved in the analysis of the network. Matrix equations describing linear and nonlinear elements of the network and those describing network topology are formulated in terms of vectors whose entries are Fourier series coefficients or time samples. These nonlinear equations are solved using numerical iteration algorithms.

In most microwave nonlinear networks we are interested in the steady-state response with periodic excitation and periodic responses with a limited number of significant harmonic frequencies. From the above considerations it is evident that the balanced methods are very well suited for microwave nonlinear network analysis.

9.1 LINEAR AND NONLINEAR SUBNETWORKS

Nonlinear analysis starts with the formulation of a circuit model to be analyzed. In microwave networks operating in large signal mode, semiconductor devices such as MESFETs or Schottky diodes create the nonlinearities. Because the rest of the network is linear, it is desirable to divide the network into a linear subnetwork and a nonlinear subnetwork that may be treated separately. Figure 9.1 presents schematically linear and nonlinear parts of a network. The package equivalent circuit as

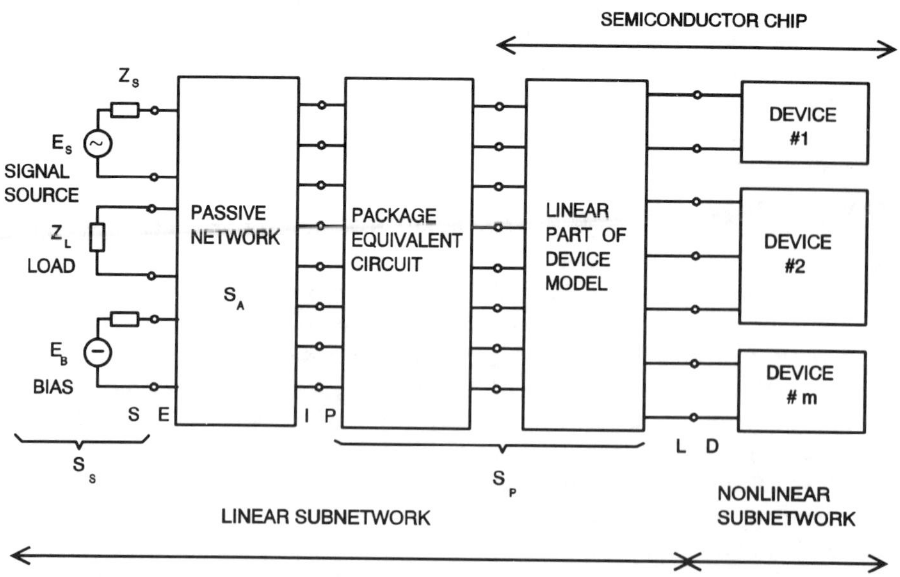

Figure 9.1 Separation of a nonlinear microwave network into a linear subnetwork and a nonlinear subnetwork.

well as linear parts of the semiconductor device models belong to the linear subnetwork.

The nonlinear circuit elements in the device models such as nonlinear current sources, voltage-dependent capacitors, voltage-controlled inductors, and current-dependent resistors create the nonlinear part of the network, which are usually considered in the time domain and are described by sets of ordinary differential equations that can be represented in the form [1]

$$\mathbf{i}(t) = \mathbf{f}\left\{\mathbf{i}(t), \frac{d\mathbf{i}}{dt}, \mathbf{v}(t), \frac{d\mathbf{v}}{dt}\right\} \tag{9.1}$$

where $\mathbf{f}$ is a vector whose elements are nonlinear functions of various currents, voltages, and their time derivatives. The form of the vector function $\mathbf{f}$ comes from the nonlinear modeling of semiconductor devices.

The linear subnetwork, on the other hand, can be described by the scattering matrices $\mathbf{S}_\mathrm{S}$, $\mathbf{S}_\mathrm{A}$, and $\mathbf{S}_\mathrm{P}$ defined in the frequency domain. According to Figure 9.1, $\mathbf{S}_\mathrm{S}$ is the scattering matrix of loads and sources exciting the analyzed network, $\mathbf{S}_\mathrm{A}$ is the scattering matrix of the passive part of the network and $\mathbf{S}_\mathrm{P}$ represents the package equivalent circuit together with linear part of the device models.

9.2 PARAMETRIC DESCRIPTION OF NONLINEAR DEVICES

The dependence between the current and the voltage of a nonlinear device may usually be given through a variable that is external to the device [6, 7]. Such external parameterization may be interpreted as a linear periodic voltage source with the internal impedance $Z_g(\omega)$ connected to the device poles. Figure 9.2 illustrates the circuit representation of the nonlinear device external parameterization by electromotive force e_g(t) of a linear periodic voltage source.

The periodic steady state in this circuit may be found by integrating differential equations describing the nonlinear device. For example, in the case of a semiconductor diode with an equivalent circuit given in Figure 9.3, the set of equations

$$\frac{dv_b}{dt} = \frac{i_d - i_b(v_b)}{C_b(v_b)} \tag{9.2}$$

$$e_g = v_b + [r_u(V_b, i_d) + R]\, i_d \tag{9.3}$$

is solved and evenly spread samples of variables v_b, i_d, i_b, and e_g are used to compute spectra V_d, I_d, and E_g of, respectively, v_d, i_d, and e_g using Fast Fourier Transform (FFT). As a result we find a pair of functions $V_d(E_g)$ and $I_d(E_g)$ that is interrelated by the linear source equation

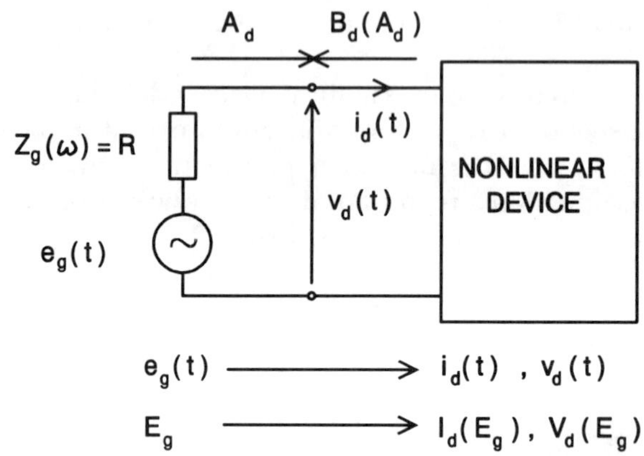

Figure 9.2 Circuit representation of a nonlinear device parameterization. Electromotive force $e_g(t)$ of a linear periodic voltage source is an external parameter of the nonlinear device.

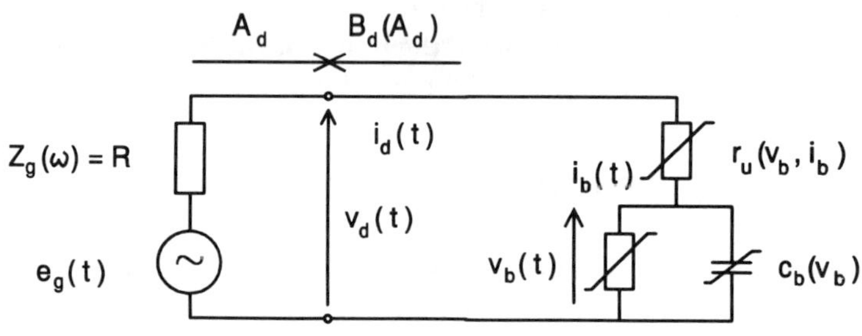

Figure 9.3 Equivalent nonlinear circuit model of a semiconductor diode junction.

$$E_g = V_d(E_g) + R\,I_d(E_g) \tag{9.4}$$

Nonlinear device parameterization may also be considered in the wave domain. Following the discussion presented in Section 2.1, (9.4) may be used to define normalized incident and reflected waves in the forms ($R > 0$)

$$A_d = \frac{1}{2\sqrt{R}}\,[V_d(E_g) + R\,I_d(E_g)] = \frac{1}{2\sqrt{R}}\,E_g \tag{9.5}$$

$$B_d = \frac{1}{2\sqrt{R}}\,[V_d(E_g) - R\,I_d(E_g)] \tag{9.6}$$

Assuming that the pair of functions $V_d(E_g)$ and $I_d(E_g)$ is known, a function $B_d(A_d)$ may be found (from (9.5) $E_g = 2\sqrt{R}\,A_d$).

9.3 HARMONIC BALANCE METHOD

The principle of the harmonic balance method comes from the observation that the periodic steady state of nonlinear components is determined by global properties of the linear subnetwork seen by the components at discrete harmonic frequencies $\{ h\,\omega_o, h = 0, 1, 2, \ldots \}$. The linear subnetwork must be characterized only at the harmonic frequencies. The harmonic balance method may be divided into three consecutive steps [8].

At the first step the linear subnetwork is reduced to the *M*-port seen from the poles of *M* nonlinear components. The scattering matrix description of the *M*-port network is the most convenient for these network transformations.

At the second stage of the analysis we determine the steady-state voltages and currents at poles of nonlinear components of the network. During this stage of analysis the linear subnetwork is represented by a reduced multiport with the number of ports equal to the number of nonlinear components of the analyzed network. Hence, the number of linear equations and the number of variables are reduced to minimums.

At the third stage, steady-state currents and voltages determined at poles of nonlinear components are used to find incident and reflected waves at the output ports of the network. This stage of analysis leads to the determination of large signal properties of the analyzed network.

In nonlinear steady-state analysis, periodic signal may be represented by one of two forms of the complex Fourier series [9]

$$x(t) = \sum_{h=-H}^{H} X^{\mathrm{d}}(h)\exp(j\,h\,\omega_o t) \tag{9.7}$$

or

$$x(t) = X^{\mathrm{s}}(0) + \sqrt{2}\,\mathrm{Re}\left\{ \sum_{h=1}^{H} X^{\mathrm{s}}(h)\exp(j\,h\,\omega_o t) \right\} \tag{9.8}$$

where $H > 0$ is the number of considered harmonics.

Coefficients of these series, which are also respectively known as double-sided and single-sided spectrum, are given by the relations

$$X^{\mathrm{d}}(h) = \frac{1}{T}\int_0^T x(t)\exp(-j\,h\,\omega_o\,t) \qquad X^{\mathrm{d}}(-h) = X^{\mathrm{d}}(h)^* \tag{9.9}$$

$$X^{\mathrm{s}}(0) = X^{\mathrm{d}}(0) \qquad X^{\mathrm{s}}(h) = \sqrt{2}\,X^{\mathrm{d}}(h) \tag{9.10}$$

It is very convenient for further considerations to write these coefficients in the form of complex vectors as

$$\mathbf{X}^{d} = \begin{bmatrix} X^{d}(H) \\ X^{d}(H-1) \\ \vdots \\ X^{d}(h) \\ \vdots \\ X^{d}(-H+1) \\ X^{d}(-H) \end{bmatrix} \qquad \mathbf{X}^{s} = \begin{bmatrix} X^{s}(0) \\ X^{s}(1) \\ \vdots \\ X^{s}(h) \\ \vdots \\ X^{s}(H) \end{bmatrix} \tag{9.11}$$

The mean power of the signal is the square of its rms value and, according to the Parseval theorem, is equal to the square of Euclidean norm of signal spectrum

$$P = X_{rms}^{2} = \frac{1}{T}\int_{0}^{T} x^{2}(t)\, dt = \|\mathbf{X}\|^{2} = \sum_{h=0}^{H} |X^{s}(h)|^{2} \tag{9.12}$$

9.3.1 Reduced Linear Subnetwork

The linear subnetwork of the network presented in Figure 9.1 can be described by a set of frequency domain linear equations considered at each harmonic $h\omega_o$, $h = 1, 2, 3, \ldots, H$, such as

$$\begin{bmatrix} \mathbf{B}_S(h) \\ \mathbf{B}_E(h) \\ \mathbf{B}_I(h) \\ \mathbf{B}_P(h) \\ \mathbf{B}_L(h) \end{bmatrix} = \begin{bmatrix} \mathbf{S}_S & \mathbf{O} & \mathbf{O} & \mathbf{O} & \mathbf{O} \\ \mathbf{O} & \mathbf{S}_{EE} & \mathbf{S}_{EI} & \mathbf{O} & \mathbf{O} \\ \mathbf{O} & \mathbf{S}_{IE} & \mathbf{S}_{II} & \mathbf{O} & \mathbf{O} \\ \mathbf{O} & \mathbf{O} & \mathbf{O} & \mathbf{S}_{PP} & \mathbf{S}_{PL} \\ \mathbf{O} & \mathbf{O} & \mathbf{O} & \mathbf{S}_{LP} & \mathbf{S}_{LL} \end{bmatrix} \begin{bmatrix} \mathbf{A}_S(h) \\ \mathbf{A}_E(h) \\ \mathbf{A}_S(h) \\ \mathbf{A}_P(h) \\ \mathbf{A}_L(h) \end{bmatrix} + \begin{bmatrix} \mathbf{C}_S(h) \\ \mathbf{O} \\ \mathbf{O} \\ \mathbf{O} \\ \mathbf{O} \end{bmatrix} \tag{9.13}$$

For convenience, spectrum components of ingoing, outgoing, and impressed waves corresponding to particular port sets of the network are grouped in the form of column vectors as

$$\mathbf{A}_p(h) = \begin{bmatrix} A_{p1}(h) \\ A_{p1}(h) \\ \vdots \end{bmatrix} \qquad \mathbf{B}_p(h) = \begin{bmatrix} B_{p1}(h) \\ B_{p1}(h) \\ \vdots \end{bmatrix} \qquad \mathbf{C}_p(h) = \begin{bmatrix} C_{p1}(h) \\ C_{p1}(h) \\ \vdots \end{bmatrix} \tag{9.14}$$

where $P \in \{ S, E, I, P, L \}$ denotes a symbol of a particular port group.

A signal source exciting the network, DC bias sources, and loads are represented by the impressed wave vector $\mathbf{C}_S(h)$ and the diagonal scattering matrix $\mathbf{S}_S$

$$\mathbf{S}_S = \begin{bmatrix} S_{S1}(h) & 0 & 0 & \cdots \\ 0 & S_{S2}(h) & 0 & \cdots \\ 0 & 0 & & \\ \vdots & \vdots & & \ddots \end{bmatrix} \tag{9.15}$$

The linear subnetwork is described by the two scattering matrices

$$\mathbf{S}_A = \begin{bmatrix} \mathbf{S}_{EE} & \mathbf{S}_{EI} \\ \mathbf{S}_{IE} & \mathbf{S}_{II} \end{bmatrix} \quad \text{and} \quad \mathbf{S}_P = \begin{bmatrix} \mathbf{S}_{PP} & \mathbf{S}_{PL} \\ \mathbf{S}_{LP} & \mathbf{S}_{LL} \end{bmatrix} \tag{9.16}$$

Taking into account connections between ports inside the linear subnetwork, we write constraints on vectors of ingoing and outgoing waves of the adjacent ports in the form

$$\begin{bmatrix} \mathbf{B}_S(h) \\ \mathbf{B}_E(h) \\ \mathbf{B}_I(h) \\ \mathbf{B}_P(h) \end{bmatrix} = \mathbf{\Gamma}_R \begin{bmatrix} \mathbf{A}_S(h) \\ \mathbf{A}_E(h) \\ \mathbf{A}_I(h) \\ \mathbf{A}_P(h) \end{bmatrix} = \begin{bmatrix} \mathbf{S}_S & \mathbf{O} & \mathbf{O} & \mathbf{O} & \mathbf{O} \\ \mathbf{O} & \mathbf{S}_{EE} & \mathbf{S}_{EI} & \mathbf{O} & \mathbf{O} \\ \mathbf{O} & \mathbf{S}_{IE} & \mathbf{S}_{II} & \mathbf{O} & \mathbf{O} \\ \mathbf{O} & \mathbf{O} & \mathbf{O} & \mathbf{S}_{PP} & \mathbf{S}_{PL} \end{bmatrix} \begin{bmatrix} \mathbf{A}_S(h) \\ \mathbf{A}_E(h) \\ \mathbf{A}_I(h) \\ \mathbf{A}_P(h) \end{bmatrix} \tag{9.17}$$

Reducing from (9.13) and (9.17) all the waves related to the internal ports of the linear subnetwork, we receive a set of equations that describes the reduced linear multiport connected to the nonlinear device ports as

$$\mathbf{B}_L(h\omega_o) = \mathbf{S}_L(h\omega_o) + \mathbf{C}_L(h\omega_o) \qquad h = 0, 1, 2 \ldots \tag{9.18}$$

where

$$\mathbf{S}_L = \mathbf{S}_{LL} + \mathbf{S}_{LR}\, W_R^{-1}\, \mathbf{S}_{RL} \tag{9.19}$$

and

$$\mathbf{C}_L = \mathbf{S}_{LR}\, W_R^{-1}\, \mathbf{C}_R \tag{9.20}$$

In (9.19) and (9.20)

$$\mathbf{W}_R = \Gamma_R - \begin{bmatrix} \mathbf{S}_S & \mathbf{O} & \mathbf{O} & \mathbf{O} \\ \mathbf{O} & \mathbf{S}_{EE} & \mathbf{S}_{EI} & \mathbf{O} \\ \mathbf{O} & \mathbf{S}_{IE} & \mathbf{S}_{II} & \mathbf{O} \\ \mathbf{O} & \mathbf{O} & \mathbf{O} & \mathbf{S}_{PP} \end{bmatrix} \qquad \mathbf{S}_{RL} = \begin{bmatrix} \mathbf{O} \\ \mathbf{O} \\ \mathbf{O} \\ \mathbf{S}_{PL} \end{bmatrix} \tag{9.21}$$

and

$$\mathbf{C}_R = \begin{bmatrix} \mathbf{C}_S \\ \mathbf{O} \\ \mathbf{O} \\ \mathbf{O} \end{bmatrix} \qquad \mathbf{S}_{LR} = [\mathbf{O} \quad \mathbf{O} \quad \mathbf{O} \quad \mathbf{S}_{LP}] \tag{9.22}$$

Figure 9.4 illustrates the reduced linear multiport connected to the nonlinear devices. An appropriate method, as discussed in Section 9.3.3, may be used to find the steady-state solution of the analyzed network.

The steady-state solution of this reduced multiport is the spectrum component vector $\mathbf{A}_L(h)$ that is determined at each harmonic $h\omega_o$. The steady-state solution spectrum component vectors $\mathbf{A}_L(h)$ are then used to find harmonic waves $\mathbf{A}_L(h)$ at

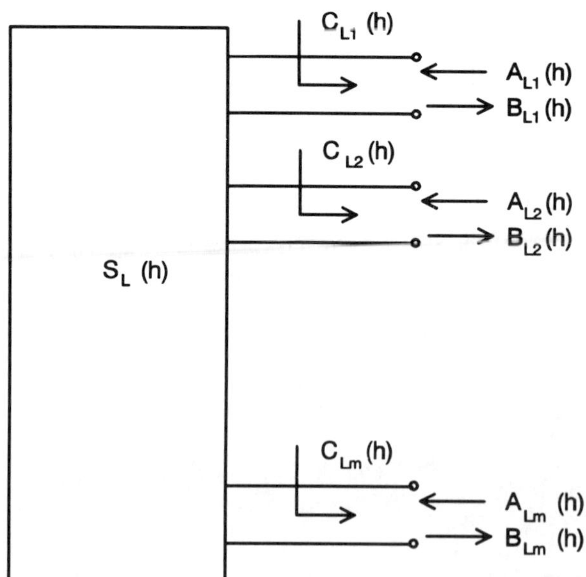

Figure 9.4 Reduced linear multiport subnetwork seen by nonlinear devices of the analyzed nonlinear microwave network.

all the network ports and, in particular, at the input and output ports grouped in $\{ S \}$. We have

$$\begin{bmatrix} \mathbf{A}_S(h) \\ \mathbf{A}_E(h) \\ \mathbf{A}_I(h) \\ \mathbf{A}_P(h) \end{bmatrix} = \mathbf{W}_R^{-1} (\mathbf{S}_{RL} \mathbf{A}_L + \mathbf{C}_D) \tag{9.23}$$

and

$$\begin{bmatrix} \mathbf{B}_S(h) \\ \mathbf{B}_E(h) \\ \mathbf{B}_I(h) \\ \mathbf{B}_P(h) \end{bmatrix} = \mathbf{\Gamma}_R \begin{bmatrix} \mathbf{A}_S(h) \\ \mathbf{A}_E(h) \\ \mathbf{A}_I(h) \\ \mathbf{A}_P(h) \end{bmatrix} \tag{9.24}$$

Computing the harmonic waves at the input and output ports $\mathbf{A}_S(h)$ and $\mathbf{B}_S(h)$ facilitates computing all the desired network functions.

Because the signal source available power at the basic frequency ω_o equals $P_{SA} = |C_{S1}(1)|^2/(1 - |S_{S1}(1)|^2)$ and because the real power delivered to the load at the hth harmonic is $P_L(h) = |A_{S2}(h)|^2 - |B_{S2}(h)|^2$, we have the *transducer power gain*

$$G_T = \frac{|A_{S2}(1)|^2(1 - |S_{S1}(1)|^2)\,(1 - |S_{S2}(1)|^2)}{|C_{S1}(1)|^2} \tag{9.25}$$

the *input port large signal reflection coefficient*

$$\Gamma_{IN} = \frac{B_{E1}(1)}{A_{E1}(1)} \tag{9.26}$$

and the *power efficiency of n times multiplier*

$$\eta_n = \frac{|A_{S2}(n)|^2(1 - |S_{S1}(1)|^2)\,(1 - |S_{S2}(n)|^2)}{|C_{S1}(1)|^2} \tag{9.27}$$

9.3.2 Balance Equation

The so-called piecewise harmonic balance technique used for the analysis of microwave nonlinear networks is based on the fact that the periodic steady state in the nonlinear part of the network depends on properties of the linear part of the network seen from the poles of the nonlinear devices at harmonic frequencies [8]. According to (9.11) the linear subnetwork may be represented as a set of independent wave sources or equivalent voltage sources connected to the poles of nonlinear devices. Figure 9.5 illustrates this circuit.

Equivalent to (9.18), the matrix equation that describes the linear subnetwork is

$$\mathbf{V}_L = \mathbf{Z}_L \, \mathbf{I}_L + \mathbf{E}_L \tag{9.28}$$

The second part of the network, the nonlinear devices, are described by a pair of vector functions $\mathbf{V}_d(\mathbf{X})$ and $\mathbf{I}_d(\mathbf{X})$, which were discussed in Section 9.2. These functions define the properties of nonlinear devices in the periodic steady state. According to the first Kirchhoff law the currents in the circuit must satisfy the equations

$$\mathbf{I}_L = - \, \mathbf{I}_d(\mathbf{X}) \tag{9.29}$$

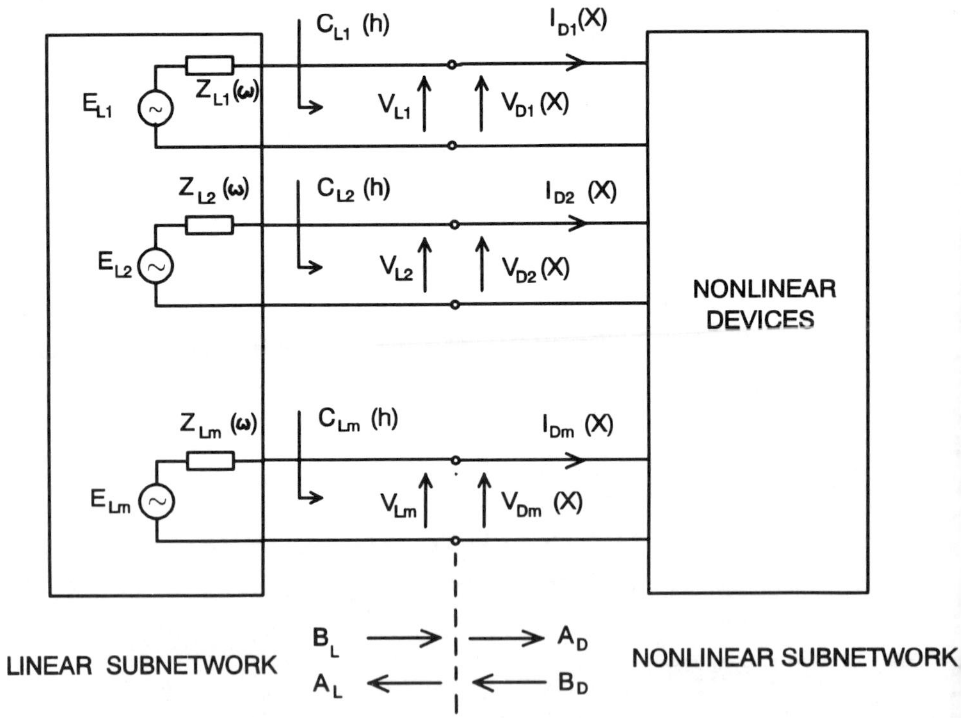

Figure 9.5 Circuit representation of the harmonic balance equation.

Then the voltages at the linear part of the network are

$$\mathbf{V}_{\mathrm{L}} = \mathbf{E}_{\mathrm{L}} - \mathbf{Z}_{\mathrm{L}}\, \mathbf{I}_{\mathrm{d}}(\mathbf{X}) \tag{9.30}$$

Voltages in the network satisfy the second Kirchhoff law. In the steady state we have

$$\mathbf{V}_{\mathrm{d}}(\mathbf{X}) = \mathbf{V}_{\mathrm{L}} \tag{9.31}$$

From (9.30) and (9.31) we obtain an equation called harmonic balance equation [1–6, 8]

$$\mathbf{V}_{\mathrm{d}}(\mathbf{X}) + \mathbf{Z}_{\mathrm{L}}\, \mathbf{I}_{\mathrm{d}}(\mathbf{X}) - \mathbf{E}_{\mathrm{L}} = \mathbf{0} \tag{9.32}$$

Because we know the functions $\mathbf{V}_{\mathrm{d}}(\mathbf{X})$ and $\mathbf{I}_{\mathrm{d}}(\mathbf{X})$, the solution of (9.32) is the vector of parametric variable **X,** which we determine by finding zeros of the function

$$\mathbf{F}(\mathbf{X}) = \mathbf{V}(\mathbf{X}) = \mathbf{V}_{\mathrm{d}}(\mathbf{X}) + \mathbf{Z}_{\mathrm{L}}\, \mathbf{I}_{\mathrm{d}}(\mathbf{X}) - \mathbf{E}_{\mathrm{L}} \tag{9.33}$$

The function $\mathbf{F}(\mathbf{X}) = \mathbf{V}(\mathbf{X})$ is called the error function of harmonic balance, the meaning of which may be better explained by rewriting it in the form

$$\mathbf{V}_{\mathrm{d}}(\mathbf{X}) + \mathbf{Z}_{\mathrm{L}}\, \mathbf{I}_{\mathrm{d}}(\mathbf{X}) - [\mathbf{E}_{\mathrm{L}} + \mathbf{V}(\mathbf{X})] = \mathbf{0} \tag{9.34}$$

When we compare this equation with the harmonic balance equation (9.32) it is obvious that for each trial vector **X** there is a vector of periodic excitations

$$\mathbf{E}(\mathbf{X}) = \mathbf{E}_{\mathrm{L}} - \mathbf{V}(\mathbf{X}) \tag{9.35}$$

that fulfills the harmonic balance equation. By successive trials we search for such a parametric variable vector **X** at which the difference between $\mathbf{E}(\mathbf{X})$ computed for the nonlinear subnetwork and $\mathbf{E}_{\mathrm{L}}$ computed as a vector of the linear subnetwork equivalent voltage sources reaches zero.

The error function of harmonic balance may be also defined in terms of a scattering matrix and wave variables. Using (9.5), (9.6), (9.18), and (9.32) we receive the error function of the harmonic balance in the form

$$\mathbf{F}(\mathbf{X}) = \mathbf{A}(\mathbf{X}) = \frac{1}{2\sqrt{R}}(\mathbf{1} - \mathbf{S}_{\mathrm{L}})\, \mathbf{V}_{\mathrm{d}}(\mathbf{X}) + \frac{\sqrt{R}}{2}(\mathbf{1} + \mathbf{S}_{\mathrm{L}})\, \mathbf{I}_{\mathrm{d}}(\mathbf{X}) - \mathbf{C}_{\mathrm{L}} \tag{9.36}$$

In (9.36) **I** is the identity matrix and R is the reference impedance common to $\mathbf{S}_L$ and to waves $\mathbf{A}_d$ and $\mathbf{B}_d$ (see (9.5) and (9.6)) used for the parametric description of nonlinear devices.

If the incident waves $\mathbf{A}_d$ are used as parametric variables, then according to (9.36) the error function is given by

$$\mathbf{F}(\mathbf{A}_d) = \mathbf{A}_d(\mathbf{A}_d) = \frac{1}{2\sqrt{R}}(\mathbf{1} - \mathbf{S}_L)\,\mathbf{V}_d(\mathbf{A}_d)$$

$$+\frac{\sqrt{R}}{2}(\mathbf{1} + \mathbf{S}_L)\,\mathbf{I}_d(\mathbf{X}) - \mathbf{C}_L$$

$$= \mathbf{A}_d - \mathbf{S}_L\,\mathbf{B}_L\,(\mathbf{A}_d) - \mathbf{C}_L \tag{9.37}$$

The error function $\mathbf{A}(\mathbf{A}_d)$ is directly related to the incident waves on the nonlinear device ports. Thanks to the wave variables, the error function of the form (9.37) is a natural representation of the interaction between the linear part and the nonlinear part of the analyzed network.

The solution of the system of equations defined by the error function of harmonic balance (see (9.33) or (9.36)) may be realized using one of many methods for the iterative solution of a set of nonlinear equations with several variables [10, 11]. All these methods generate successive estimates of the solution vector $\mathbf{X}$ of the equation $\mathbf{F}(\mathbf{X}) = O$ in such a way that a selected norm of the error function decreases with increases in the number of iterations. The error function norm is usually the Euclidean norm. In the case of the harmonic balance error function in the wave form (9.37), this norm is equal to the ratio of rms value of the error and rms value of magnitudes of the impressed waves of the equivalent wave source; that is,

$$\varepsilon(\mathbf{A}_d^{(n)}) = \frac{||\mathbf{A}(\mathbf{A}_d^{(n)})||}{||\mathbf{C}_L||} \tag{9.38}$$

where n is the iteration number.

Because nonlinear devices are usually described by partial differential equations with voltages and currents as dependent variables, nonlinear analysis methods cannot be formulated using only wave variables. At some state of the analysis, waves at nonlinear device ports must be transformed into device pole voltages and currents.

9.3.3 Multireflection Method

The multireflection method belongs to a class of relaxation methods used to solve the harmonic balance equation [10, 12, 13]. This method is a peculiar one among the other known methods because it has a very clear circuit representation and because the multireflection method has proved to be very reliable.

Figure 9.6 illustrates a modified version of the analyzed network in which sections of lossless transmission lines are inserted between ports of the nonlinear devices and corresponding ports of the linear subnetwork. These transmission lines create the basis for the method. It is assumed that the characteristic impedances of the lines are real and equal to R_o and that the lengths of lines are integer multipliers of the wave length λ_o in the transmission lines excited at frequency ω_o. If we assume now that the circuit with transmission lines excited at frequency ω_o has reached its steady state, this state is also a steady state of the original network (without transmission lines between subnetworks) because the transmission lines are transparent at all frequencies $h\omega_o$, $h = 1, 2, \ldots, H$.

The numerical algorithm for finding the steady state in the original network may be simulated then by multireflections in the network with long lossless transmission lines. It must be assumed that the propagation time $t_d = 2\Pi\, 1/\omega_o$ in transmission lines is much longer than the response times of nonlinear devices and the linear subnetwork.

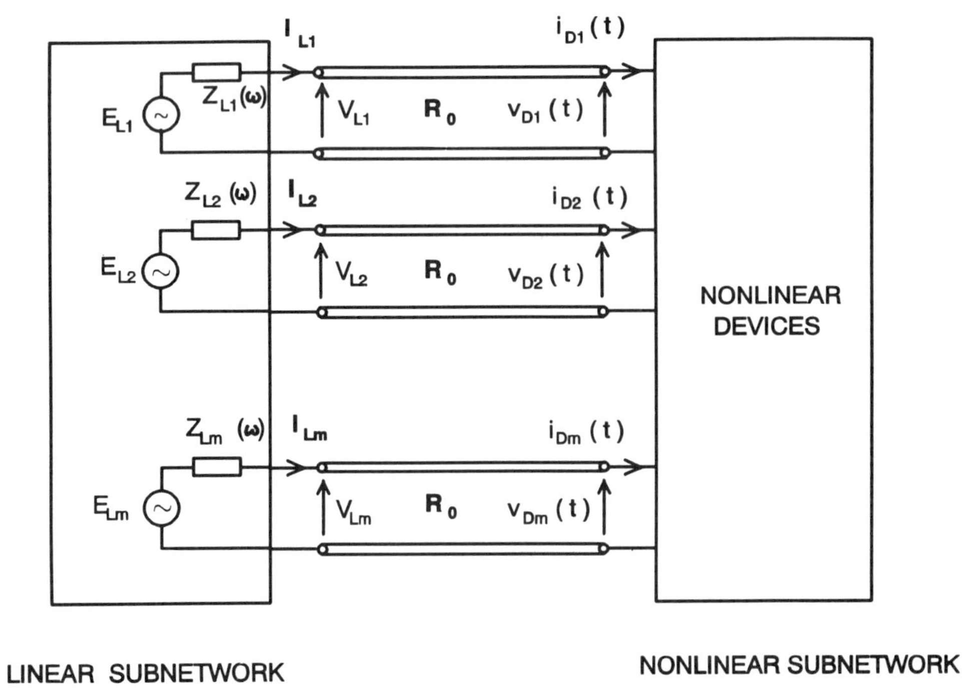

Figure 9.6 General nonlinear network with fictitious ideal transmission lines inserted between the linear and the nonlinear subnetworks.

When sinusoidal sources $e_{Lm}(t)$ are originally switched on, incident waves are impressed on the transmission lines. These waves propagate along the lines and, after propagation time t_d, reach the respective ports of the nonlinear devices. The waves excite the nonlinear devices, and reverse traveling waves are generated. At the linear subnetwork side port of the transmission lines these waves are again reflected from $Z_L(\omega)$ and add to the incident waves. Then the process is repeated. These new incident waves travel toward the nonlinear device ports, excite the devices, and are reflected back, modifying the reflected waves. In this multireflection iteration process steady-state conditions are eventually reached. This is indicated by minimal changes in the incident waves between iterations.

Let us now consider a transmission line presented in Figure 9.7. At the instant the signal source is turned on, the input impedance of the transmission line is equal to its characteristic impedance R_o because no reflected wave exists in the line. The original incident wave at the left-hand side port $v_+^{(0)}(t)$ is found by simple analysis of the linear subnetwork. When the incident wave reaches the right-hand side end of the transmission line, the nonlinear device sees the line as a voltage source shown in Figure 9.7(d). The voltage and current in the line are superpositions of voltage and current waves propagating in the line

$$v(x,t) = v_+(x,t) + v_-(x,t) \tag{9.39}$$

$$i(x,t) = \frac{v_+(x,t) - v_-(x,t)}{R_o} \tag{9.40}$$

The relation defined by these equations may be represented as two connected-in-series voltage sources each with internal impedance R_o. Such a circuit is shown in Figure 9.7(c). Voltages at the ends of the line are

$$v(0,t) = R_o\ i(0,t) + 2\ v_-(0,t) \tag{9.41}$$

$$v(1,t) = -R_o\ i(0,t) + 2\ v_+(0,t) \tag{9.42}$$

The reverse wave $v_-^{(0)}$ is found using the relation

$$2\ v_-^{(0)}(t) = v_d^{(0)} - i_d^{(0)}\ R_o \tag{9.43}$$

The voltage $v_d^{(0)}(t)$ and the current $i_d^{(0)}(t)$ in (9.42) are determined by the numerical integration of differential equations that describe the circuit in Figure 9.7(d). The differential equations of nonlinear devices were discussed in Section 9.2. As a result of integration, vectors of samples of nonlinear device voltage $v_d^{(0)}$ and current $i_d^{(0)}$ are obtained.

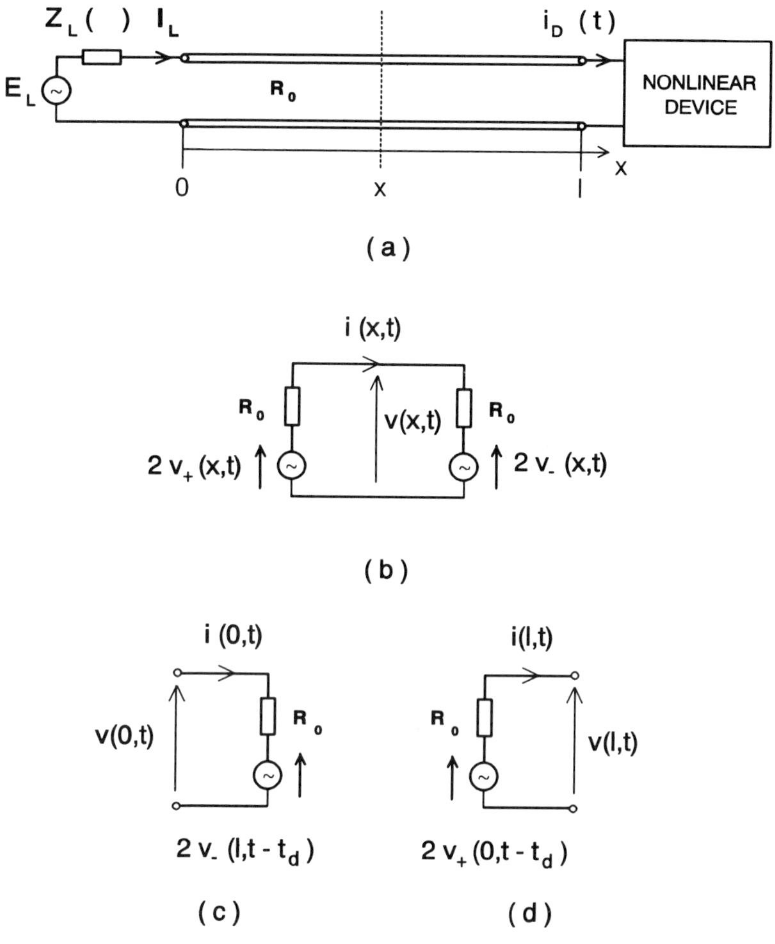

Figure 9.7 Circuit interpretation of multireflection algorithm for solving the harmonic balance equation: (a) a fictitious very long transmission line inserted between ports of linear and nonlinear subnetworks, (b) two voltage sources with internal impedances R_o represent voltage and current in the line, (c) a voltage source with internal impedance R_o representing the line at its ''linear'' left end, and (d) a voltage source with internal impedance R_o representing the line at its ''nonlinear'' right end.

To form the new incident wave, the spectrum $V_-^{(0)}$ of the reverse wave $v_-^{(0)}$ must be computed using the Fast Fourier Transform.

The new incident wave is derived as

$$2V_+^{(1)}(h) = V_L^{(1)}(h) + R_o\, I_L^{(1)}(h) \tag{9.44}$$

where the voltage $V_L^{(1)}(h)$ and the current $I_L^{(1)}(h)$ components at the input of the line are given as

$$V_L^{(1)}(h) = \frac{E_L(h)\, R_o + 2\, V_-^{(0)}(h)\, Z_L(h\omega_o)}{Z_L(h\omega_o) + R_o} \tag{9.45}$$

and

$$I_L^{(1)}(h) = \frac{E_L(h) + 2\, V_-^{(0)}(h)}{Z_L(h\omega_o) + R_o} \tag{9.46}$$

When the spectrum $V_+^{(1)}$ of the new incident wave is known, the waveform $v_+^{(1)}(t)$ may be computed using the inverse Fourier transform.

This iteration process is continued until the steady state is reached, which is indicated by stabilization of the incident wave spectrum. As a measure of the solution accuracy, the relative error in the impedances of the linear subnetwork and the nonlinear device $Z_d^{(i)}(h\omega_o) = V_d^{(i)}(h)/I_d^{(i)}(h)$ for all considered harmonics may be used in the form [12]

$$\varepsilon(h) = \left| \frac{Z_d(h\omega_o)}{Z_L(h)} \right| - 1 \tag{9.47}$$

The multireflection method is reliable and easy to implement. Its simplicity and flexibility make it particularly suitable for investigating new nonlinear devices.

It is also possible to incorporate thermal effects in the nonlinear network analysis. Usually it is assumed that the temperature of nonlinear devices is constant. However, in real network structures, power dissipated in nonlinear devices may increase their temperatures because of the limited thermal impedance of these elements. In the case of semiconductor devices, their parameters are strongly temperature dependent.

When the driving signal is a *continuous wave* (CW), the thermal effects can be included in the analysis by adding an external loop in which device temperatures are controlled and corrected. The harmonic balance equation is solved in the internal loop. From the steady-state solution of the harmonic balance equation, the power dissipated in the devices is determined. That power and thermal model of device structure facilitates deriving the new device temperature and its corresponding new device parameters. In the next step, the new harmonic balance equation is solved and the external loop is used to correct device temperatures. Of course, appropriate measures for the accuracy of the solution for both internal and external loops must be used to stop the iteration processes.

In the case when the driving signal is modulated, the analysis procedure is much more complicated [14]. It requires knowledge of dynamic thermal models of nonlinear devices.

References

[1] Chua, L. O., and P. Lin, *Computer-Aided Analysis of Electronic Circuits,* Englewood Cliffs, NJ: Prentice-Hall, 1975.

[2] Kundert, K. S., and A. Sangiovanni-Vincentelli, "Simulation of Nonlinear Circuits in the Frequency Domain," *IEEE Trans. Computer-Aided Design,* Vol. CAD-5, No. 4, October 1986, pp. 521–535.

[3] Kundert, K. S., J. K. White, and A. Sangiovanni-Vincentelli, *Steady-State Methods for Simulating Analog and Microwave Circuits,* Norwell, MA: Kluwer Academic Publishers, 1990.

[4] Rizzoli, V., and A. Neri, "State of the Art and Present Trends in Nonlinear Microwave CAD Technique," *IEEE Trans. Microwave Theory Tech.,* Vol. MTT-36, No. 2, February 1988, pp. 343–365.

[5] Maas, S. A., *Nonlinear Microwave Circuits,* Norwood, MA: Artech House, 1988.

[6] Faber, M. T., J. Chramiec, and M. E. Adamski, *Microwave and Millimeter-Wave Diode Frequency Multipliers,* Norwood, MA: Artech House, 1995.

[7] Adamski, M. E., Private Communications, Institute of Electronics Fundamentals, Warsaw University of Technology, Warsaw, Poland, 1995.

[8] Nakhla, M. S., and J. Vlach, "A Piecewise Harmonic-Balance Technique for Determination of Periodic Response of Nonlinear Systems," *IEEE Trans. Circuits Systems,* Vol. CAS-23, No. 2, February 1976, pp. 85–91.

[9] Papoulis, A., *Signal Analysis,* New York: McGraw-Hill Book Company, 1977.

[10] Adamski, M. E., and M. T. Faber, *Methods and Techniques for Nonlinear Analysis of Microwave Circuits with Diodes,* Scientific Reports, Institute of Electronics Fundamentals, Warsaw University of Technology, Warsaw, Poland, 1995.

[11] Ortega, J. M., and W. C. Rheinboldt, *Iterative Solution of Nonlinear Equations in Several Variables,* New York: Academic Press, 1970.

[12] Kerr, A. R., "A Technique for Determining the Local Oscillator Waveform in Microwave Mixers," *IEEE Trans. Microwave Theory Tech.,* Vol. MTT-23, No. 10, October 1975, pp. 828–831.

[13] Faber, M. T., and W. K. Gwarek, "Nonlinear–Linear Analysis of Microwave Mixer with any Number of Diodes," *IEEE Trans. Microwave Theory Tech.,* Vol. MTT-28, No. 11, November 1980, pp. 1174–1181.

[14] Rizzoli, V., A. Liparini, V. D. Esposti, F. Mastri, and C. Cecchetti, "Simultaneous Thermal and Electrical Analysis of Nonlinear Microwave Circuits," *IEEE Trans. Microwave Theory Tech.,* Vol. MTT-40, No. 7, July 1992, pp. 1446–1455.

CHAPTER 10

Noise Wave Parameter Measurements

As was shown in Chapter 4, the noise wave approach simplifies active device modeling.

The conventional method used for measurements of standard active two-port noise parameters F_{min}, Γ_{opt}, and R_n are based on repeated noise factor measurements for various values of the signal source reflection coefficient presented to the input port of the active *device under test* (DUT) [1–3]. This method requires a source-pull tuner, which significantly complicates the measurement procedure. The drawbacks of this method are the restricted range of measurement frequency, the limited range of Γ_S changes, a complicated calibration procedure, and problems with measurement repeatability. Tuners that are used in the measurement apparatus are complicated and very expensive mechanical devices that make automation of the measurement procedure difficult unless electronically variable varactor tuners, which have limited tuning as well as frequency range of operation, are used. Another problem encountered with this method is the fact that it requires large values of the reflection coefficient to be presented to the input port of the DUT. It is observed, in practice, that large values of Γ_S can lead to instability of a transistor under test. Meaningful measurements for a transistor that is instable are not possible.

It is clear that with $\Gamma_S = 0$ the device under test (DUT) will usually be stable. Furthermore, it was shown in Section 4.6, that under condition $\Gamma_S = 0$, the standard noise parameters F_{min}, Γ_{opt}, and R_n can be written in terms of noise wave parameters of the device [4]. These interrelations are given by (4.100) to (4.103). All this means that all the difficulties related to the tuner method can be overcome by using a noise wave parameter analyzer to directly measure noise wave parameters of the device under test. Because the input and the output ports of the DUT is loaded with matched terminations, the system is very stable in its operation.

10.1 MICROWAVE SIX-PORT NOISE WAVE PARAMETER ANALYZER

Figure 10.1 shows a system for measuring the noise wave parameters of an electronic device under test [5, 6]. The DUT is represented as a two-port with scattering matrix **S** emanating from its ports noise waves c_1 and c_2. Two circulators located at each port of the device inject noise from the the noise source into the device. Noise power that comes from the amplifiers A_1 and A_2 are terminated by matched noise sources, which prevents correlated noise power generated by the amplifiers from masking the correlated noise power generated by the DUT.

The noise waves c_1 and c_2 traveling away from the input and output ports of the DUT are combined with scattered noise waves from noise sources and produce noise waves d_1 and d_2. These noise waves are then amplified and directed to two input ports of the six-port noise wave correlation network. Two amplifiers located at the input ports of the noise correlation network increase levels of noise signals that pass through the network. This increases the sensitivity of the measurement system. At each of the four output ports of the six-port correlation network, power level measurements of noise waves e_1, e_2, e_3, and e_4 are made by a noise power spectral density meter. A microwave mixer and an associated local oscillator convert the noise

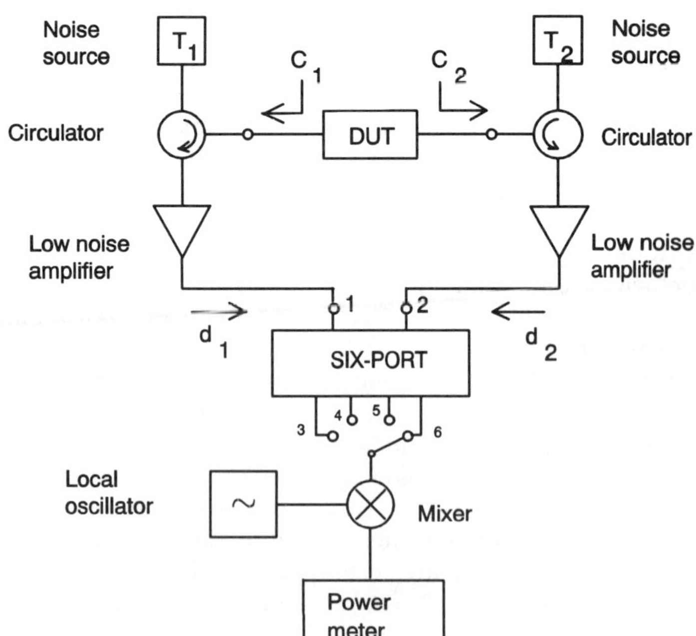

Figure 10.1 Six-port noise wave parameter analyzer. Two noise waves c_1 and c_2 represent noise generated in the device under test (DUT).

power from various output ports of the six-port correlation network to the lower intermediate frequencies that are measured by the power spectral density meter.

Figure 10.2(a) presents a configuration of the six-port noise correlation network. It is composed of four interconnected basic modules: three 180° hybrids and one quadrature (90°) hybrid [7]. These two types of hybrid directional couplers and the relationships that exist among the incident and outgoing waves are shown in Figure 10.2(b). It is assumed here that these hybrids are lossless and matched at all ports.

Incident noise waves at the input ports of the six-port correlation network designed as d_1 and d_2 are

$$d_1 = c_1 + \sqrt{kT_1}\, S_{11} + \sqrt{kT_2}\, S_{12} \tag{10.1}$$

$$d_2 = c_2 + \sqrt{kT_1}\, S_{21} + \sqrt{kT_2}\, S_{22} \tag{10.2}$$

In the above equations k is Boltzmann's constant and T_1 and T_2 are the equivalent temperature of the two noise sources, each capable of taking on two known cold and hot values.

Quantities e_1, e_2, e_3, and e_4 represent noise waves at the output ports of the six-port correlation network. With the help of Figure 10.2(a) it can be shown that the amplitudes of these noise waves are

$$e_1 = \frac{1}{\sqrt{2}}\left(d_1 - j\frac{d_2}{\sqrt{2}}\right) \tag{10.3}$$

$$e_2 = -j\frac{1}{2}\left[d_1 + (1+j)\frac{d_2}{\sqrt{2}}\right] \tag{10.4}$$

$$e_3 = -j\frac{1}{2}\left[d_1 - (1-j)\frac{d_2}{\sqrt{2}}\right] \tag{10.5}$$

$$e_4 = -j\frac{d_2}{2} \tag{10.6}$$

Using relations (10.3) to (10.6) it can be shown that power readings P_1, P_2, P_3, and P_4 are related to the noise measurable quantities $\overline{|d_1|^2}$, $\overline{|d_2|^2}$, and $\overline{d_1 d_2^*}$ by the equations

$$\overline{|d_1|^2} = P_1 + P_2 + P_3 - 3\,P_4 \tag{10.7}$$

$$\overline{|d_2|^2} = 4\,P_4 \tag{10.8}$$

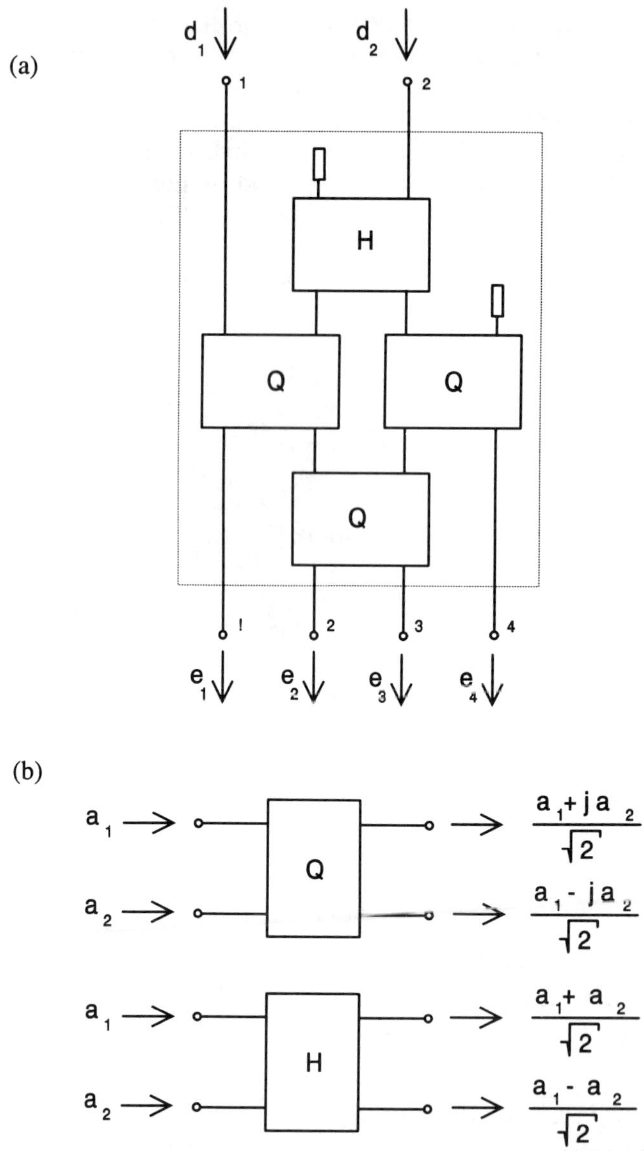

Figure 10.2 (a) Passive six-port correlation network composed of three quadrature hybrids denoted as Q (90° hybrid) and one 180° hybrid, H. (b) The basic modules of the six-port network and relations for signals: Q, quadrature hybrid, H, 180° hybrid.

$$\mathrm{Re}\{\overline{d_1 d_2^*}\} = \sqrt{2}\,(P_2 - P_3) \tag{10.9}$$

$$\mathrm{Im}\{\overline{d_1 d_2^*}\} = \frac{1}{\sqrt{2}}\,(P_2 - P_1 + P_3 - P_4) \tag{10.10}$$

where, according to (10.1) and (10.2),

$$\overline{|d_1|^2} = \overline{|c_1|^2} + kT_1\,|S_{11}|^2 + kT_2|S_{21}|^2 \tag{10.11}$$

$$\overline{|d_2|^2} = \overline{|c_2|^2} + kT_1\,|S_{21}|^2 + kT_2|S_{22}|^2 \tag{10.12}$$

$$\overline{d_1 d_2^*} = \overline{c_1 c_2^*} + kT_1\,S_{11}S_{21}^* + kT_2\,S_{12}S_{22}^* \tag{10.13}$$

Relations (10.7) to (10.13) represent a set of linear equations with twelve unknowns: $\overline{|c_1|^2}$, $\overline{|c_2|^2}$, $\mathrm{Re}\{\overline{c_1 c_2^*}\}$, $\mathrm{Im}\{\overline{c_1 c_2^*}\}$, $\mathrm{Re}\{S_{11}S_{21}^*\}$, $\mathrm{Im}\{S_{11}S_{21}^*\}$, $\mathrm{Re}\{S_{12}S_{22}^*\}$, and $\mathrm{Im}\{S_{12}S_{22}^*\}$, of which the first four are noise wave parameters of the device under test. By making power measurements at hot and cold values of T_1 and T_2 (the temperatures of noise sources) and then solving the above system of equations, it is possible to determine the noise parameters of the DUT. The set of equations under consideration is overdetermined (sixteen equations and twelve unknowns), and its solution requires statistical analysis. The values of the noise wave parameters $\overline{|c_1|^2}$, $\overline{|c_2|^2}$, and $\overline{c_1 c_2^*}$ obtained from the solution of (10.7) to (10.13) may be used to compute the standard noise parameters F_{min}, Γ_{opt}, and R_n of the two-port under test [4].

The intrinsic noise performance of the measurement system can be determined by a calibration procedure. The intrinsic noise contribution can be removed from measurements by using the embedding equation (4.67) discussed in Section 4.4.

There are other possible configurations of the measurement system [5, 6]. For example, one may use the single intermediate frequency network presented in Figure 10.3. In such a case, the noise power that originated from the DUT and from two noise sources is, by mixers, converted down to an intermediate frequency.

In both configurations of the noise wave parameter analyzer shown in Figures 10.1 and 10.3, it is possible to replace the microwave circulators by the respective quadrature hybrid directional couplers, as shown in Figure 10.4.

In this case, the noise waves d_1 and d_2 that are produced by noise waves c_1 and c_2 of the DUT and by scattered noise from noise sources are

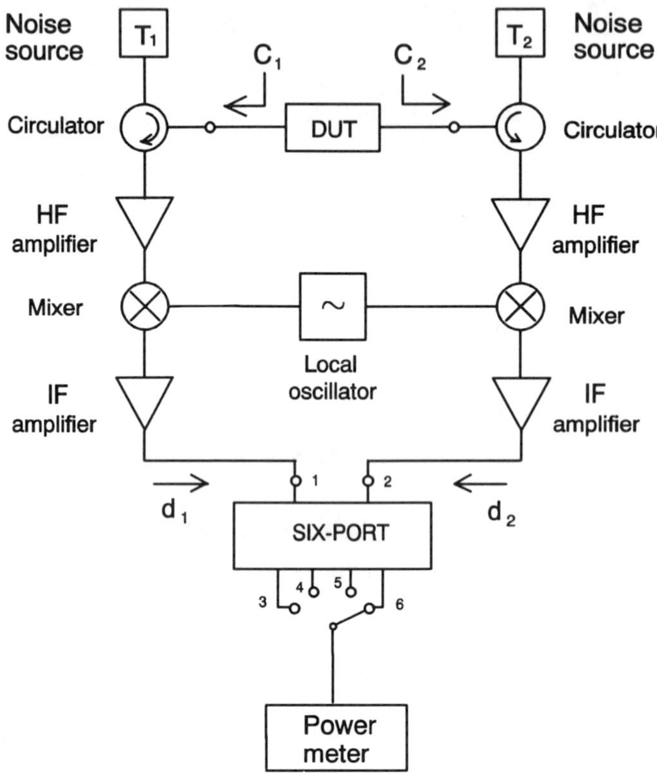

Figure 10.3 Alternative configuration of a six-port noise wave parameter analyzer with a single intermediate frequency.

$$d_1 = -\frac{j}{\sqrt{2}}\,c_1 - \frac{j}{2}\sqrt{kT_1}\,S_{11} - \frac{1}{2}\sqrt{kT_2}\,S_{12} \tag{10.14}$$

$$d_2 = +\frac{j}{\sqrt{2}}\,c_2 + \frac{1}{2}\sqrt{kT_1}\,S_{21} - \frac{j}{2}\sqrt{kT_2}\,S_{22} \tag{10.15}$$

The noise wave technique of noise parameter measurement has several advantages. It is particularly suitable for on-wafer and on-chip measurement of active device noise parameters. In general, it is very convenient for any situation where the DUT is relatively inaccessible and also when one wants to measure the noise performance of the device at cryogenic temperatures. Realizing the measurement system does not require source-pull tuner. The system may be configured successfully at low as well as at millimeter wave frequencies.

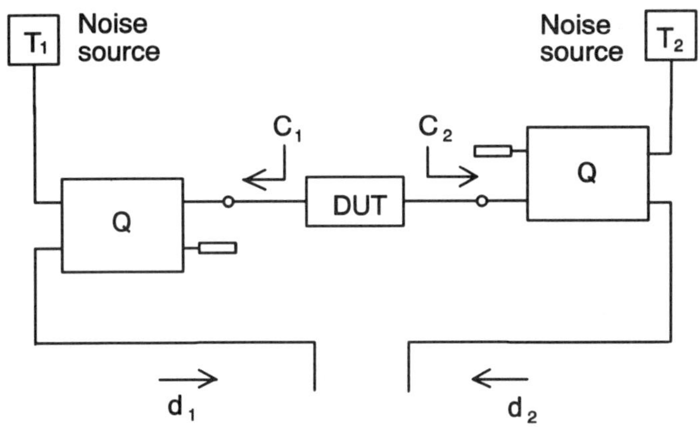

Figure 10.4 Two quadrature hybrid directional couplers replace the circulators in the alternative embodiment of the noise wave parameter analyzer.

10.2 INTERFEROMETER SYSTEM FOR NOISE WAVE PARAMETER MEASUREMENTS

Another way to determine the active two-port noise wave parameters are via interferometric measurements [8]. In an interferometer measurement system, the intermediate noise signals from two channels are added and detected. The block diagram of the interferometer system is shown in Figure 10.5.

Incident noise waves at the input ports of both channels of the interferometer are

$$d_1 = c_1 + \sqrt{kT_1}\, S_{11} + \sqrt{kT_2}\, S_{12} \tag{10.16}$$

$$d_2 = c_2 + \sqrt{kT_1}\, S_{21} + \sqrt{kT_2}\, S_{22} \tag{10.17}$$

where, according to Figure 10.5, c_1 and c_2 are noise waves traveling away from the input and output ports of the DUT.

These two noise signals are first amplified in each arm of the interferometer, and their sum $d_1 G_1 e^{j(\Phi_1+\beta)} + d_2 G_2 e^{j\Phi_2}$ and their difference $d_1 G_1 e^{j(\Phi_1+\beta)} - d_2 G_2 e^{j\Phi_2}$ are then measured by power detectors.

The measured power of the sum of the noise signals is

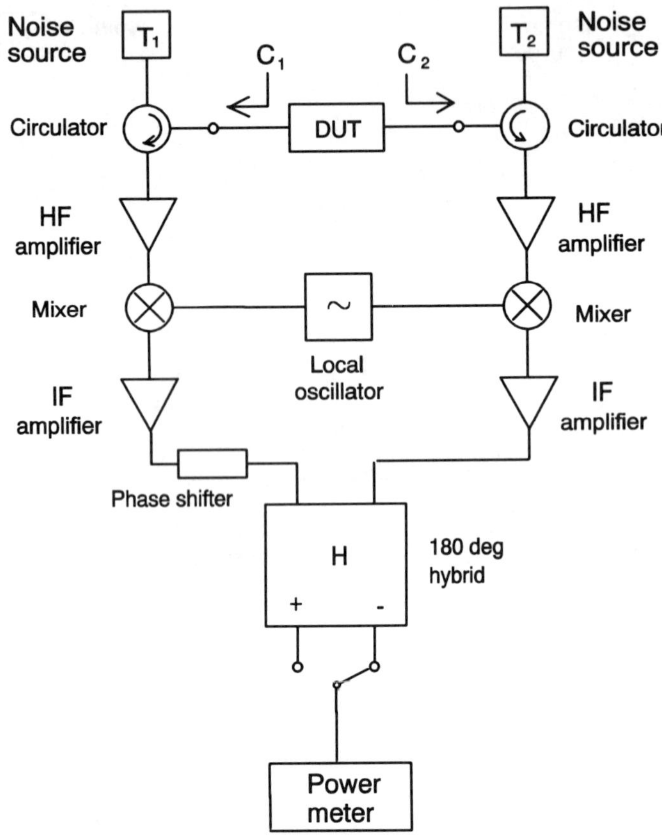

Figure 10.5 Microwave interferometer for measuring the noise wave parameters of a two-port active device.

$$
\begin{aligned}
P^+ = \overline{|c_1|^2}\, G_1^2 + \overline{|c_2|^2} \\
&+ 2\, G_1\, G_2\, \mathrm{Re}\{\overline{c_1 c_2^*}\} \cos(\Phi_1 - \Phi_2 + \beta) \\
&- 2\, G_1\, G_2\, \mathrm{Im}\{\overline{c_1 c_2^*}\} \sin(\Phi_1 - \Phi_2 + \beta) \\
&+ kT_1\, (G_1^2\, |S_{11}|^2 + G_2^2\, |S_{21}|^2) + kT_2\, (G_1^2\, |S_{12}|^2 + G_2^2\, |S_{22}|^2) \\
&+ 2\, kT_1\, G_1 G_2\, \mathrm{Re}\{S_{11} S_{21}^*\} \cos(\Phi_1 - \Phi_2 + \beta) \\
&- 2\, kT_1\, G_1 G_2\, \mathrm{Im}\{S_{11} S_{21}^*\} \sin(\Phi_1 - \Phi_2 + \beta) \\
&+ 2\, kT_2\, G_1 G_2\, \mathrm{Re}\{S_{12} S_{22}^*\} \cos(\Phi_1 - \Phi_2 + \beta) \\
&- 2\, kT_2\, G_1 G_2\, \mathrm{Im}\{S_{12} S_{22}^*\} \sin(\Phi_1 - \Phi_2 + \beta)
\end{aligned}
\tag{10.18}
$$

For the difference of the noise signals of two interferometer channels, the measured power is

$$
\begin{aligned}
P^{-} = {} & \overline{|c_1|^2}\, G_1^2 + \overline{|c_2|^2} \\
& - 2\, G_1\, G_2\, \mathrm{Re}\{\overline{c_1 c_2^*}\} \cos(\Phi_1 - \Phi_2 + \beta) \\
& + 2\, G_1\, G_2\, \mathrm{Im}\{\overline{c_1 c_2^*}\} \sin(\Phi_1 - \Phi_2 + \beta) \\
& + kT_1\, (G_1^2\, |S_{11}|^2 + G_2^2\, |S_{21}|^2) + kT_2\, (G_1^2\, |S_{12}|^2 + G_2^2\, |S_{22}|^2) \\
& - 2\, kT_1\, G_1 G_2\, \mathrm{Re}\{S_{11} S_{21}^*\} \cos(\Phi_1 - \Phi_2 + \beta) \\
& + 2\, kT_1\, G_1 G_2\, \mathrm{Im}\{S_{11} S_{21}^*\} \sin(\Phi_1 - \Phi_2 + \beta) \\
& - 2\, kT_2\, G_1 G_2\, \mathrm{Re}\{S_{12} S_{22}^*\} \cos(\Phi_1 - \Phi_2 + \beta) \\
& + 2\, kT_2\, G_1 G_2\, \mathrm{Im}\{S_{12} S_{22}^*\} \sin(\Phi_1 - \Phi_2 + \beta)
\end{aligned}
\tag{10.19}
$$

The noise wave parameters of the DUT can be determined by making measurements with hot and cold values of T_1 and T_2 of the noise sources, with each channel of the interferometer turned on and off, and with various phase shifts.

The interferometer technique is very attractive for on-wafer and on-chip noise wave parameter measurements of microwave active devices and is also very promising for the noise characterization of microwave networks at cryogenic temperatures. It does not require source-pull tuner and is attractive for millimeter wave application.

References

[1] Lane, R. Q., "The Determination of the Device Noise Parameters," *Proc. IEEE,* Vol. 116, August 1969, pp. 1461–1462.

[2] Mitama, M., and H. Katoh, "An Improved Computational Method for Noise Parameter Measurement," *IEEE Trans. Microwave Theory Tech.,* Vol. MTT-27, June 1979, pp. 612–615.

[3] Pospieszalski, M. W., "On the Measurement of Noise Parameters of Microwave Two-Ports," *IEEE Trans. Microwave Theory Tech.,* Vol. MTT-34, April 1986, pp. 456–458.

[4] Dobrowolski, J. A., "A CAD-Oriented Method for Noise Figure Computation of Two-Ports with Any Internal Topology," *IEEE Trans. Microwave Theory Tech.,* Vol. MTT-37, January 1989, pp. 15–20.

[5] Wedge, S. W., and D. B. Rutledge, " Wave Techniques for Noise Modeling and Measurement," *IEEE Trans. Microwave Theory Tech.,* Vol. 40, November 1992, pp. 2004–2012.

[6] Wedge, S. W., and D. B. Rutledge, *Microwave Six-Port Noise Parameter Analyser,* U.S. Patent No. 5, 170, 126, December 8, 1992.

[7] Engen, G. F., "An Improved Circuit for Implementing the Six-Port Technique of Microwave Measurements," *IEEE Trans. Microwave Theory Tech.,* Vol. MTT-26, December, 1977, pp. 1080–1083.

[8] Withington, S., "Scattering Noise Waves in Microwave and mm Wave Networks," *Microwave J.,* Vol. 32, June 1989, pp. 169–178.

APPENDIXES

Reference Manual of JADMIC2— A General-Purpose Program for Microwave Network Analysis and Design

APPENDIX A

JADMIC2 is the personal computer–based software developed for the analysis and design of linear microwave circuits. The development of the JADMIC2 software followed four objectives:

- JADMIC2 should run on PC platforms commonly found in engineering environments: 286-class, 386-class, or 486-class DOS personal computers with math co-processors, 640K of RAM, hard disk, and VGA graphics.
- JADMIC2 should be intuitive, menu driven, and easy to use.
- JADMIC2 should correctly solve realistic linear microwave circuit design problems of interest to educational, industrial, and research communities.
- JADMIC2 documentation should be very extensive and highly understandable to the user, who may want to introduce his own ideas into the software.

The theory behind the analysis and optimization procedures used in this program is described in the book.

The appendixes contain detailed information on the input file preparation, operation of JADMIC2, the output quantities, and their presentation.

APPENDIX B

JADMIC2 Operation

JADMIC2 uses the cursor keys and pull-down menus to control program execution. Generally, to escape from any operation selection level in the menu, the escape <esc> key on the keyboard is used.

A message line at the bottom of the screen provides information about the current directory in use by the program for input and output file storage.

In order to solve a problem using JADMIC2, the user must create an input file that describes the network to be analyzed and/or optimized and specifies desired output data from the program. The input data file is created by means of special commands (mnemonics) and numbers. All these statements start with self-explanatory key words that must be arranged properly into different sections (blocks). The input data file should be prepared before running JADMIC2. For this purpose the user should use one of many easy-to-use full-screen editors that are currently available on the market.

The JADMIC2 program is equipped with its own editor that can be used to create and to modify the input file before successive runs.

The results of a microwave circuit analysis and/or design are located in the output data. Several circuit performance characteristics can be computed. A full list of the available circuit functions is contained in Appendix I.

JADMIC2 also provides plotted output. All calculated responses may be presented as plots on the monitor screen using the integrated plot package.

B.1 INTERACTIVE MODE OF JADMIC2 OPERATION

The JADMIC2 commands are organized into menus; the menu names appear along the menu bar at the top of the screen. Once JADMIC2 is executed, the overall operation of the program is controlled by the pull-down menu shown in the screen shown in Figure B.1.

The main menu has five choices, which may be selected with the left and right arrow keys. Within each highlighted box of the submenu selection, there are several command choices, and these may be selected by the up and down arrow (cursor) keys.

B.1.1 The Input Menu

Execution of the JADMIC2 always starts from the Input submenu in the menu screen shown in Figure B.1.

By using the up and down arrow (cursor) keys, the user may select the desired command in the Input submenu. By pressing the <enter> key, the user invokes execution of the selected command.

All commands on the Input menu are described below in alphabetical order.

The Directory manager command displays the list of circuit and data files in the current directory. By using the arrow up and down keys and then the enter <enter> key, the user may select the desired file as the input file for JADMIC2.

The Edit circuit file command is used to access an internal JADMIC2 editor. The input file can be prepared before running JADMIC2 or can be entered and edited within JADMIC2 by using its own internal editor. It is intended to use the internal JADMIC2 editor mainly to introduce changes into the existing network file. These changes may be introduced once the circuit is analyzed. After doing so, however, it is important that the user updates the file on the disk before the analysis of the updated circuit is performed again.

The Load circuit file command loads the selected file into the JADMIC2 program.

The Exit command is used to exit from the Input menu to the main menu.

```
Input   Analysis   Optimization   Output   Help   Exit
 Load Circuit File
 Edit Circuit File
 View Error File
 Directory Manager
 Exit

C:\JADMIC\
```

Figure B.1 JADMIC2 command menu.

B.1.2 The Analysis Menu

The Analysis menu is shown in Figure B.2.

```
Analysis
  Frequency Analysis
  Exit
```

Figure B.2 The Analysis menu.

By using the up and down arrow keys, the user may select the type of analysis performed by the JADMIC2 program.

The commands on the Analysis menu screen are described below.

The Frequency analysis command starts execution of the circuit analysis in the frequency domain. The JADMIC2 generates printed and plotted output data as requested by the input file.

The Yield analysis command starts execution of the statistical analysis for element tolerances according to the tolerance specifications given within the input file.

The present version of JADMIC2 does not perform yield analysis of a network.

The Exit command is used to exit from the Analysis menu.

B.1.3 The Optimization Menu

Figure B.3 illustrates the Optimization menu of JADMIC2.

This menu permits the definition of the type of optimization performed by the JADMIC2 program. Choosing any one of these items with the up and down arrow keys and pressing the <enter> key will select the desired type of optimization procedure. The present version of JADMIC2 allows using the Powell or simplex method for optimization of a network.

```
Optimization
  Powell
  Exit
```

Figure B.3 The Optimization menu.

B.1.4 The Output Menu

Selecting the Output menu permits the selection of the form of the output data presentation. The output menu is shown in Figure B.4.

All commands on the Output menu are described below.

```
Output
  Graphics
  View Output File
  Exit
```

Figure B.4 The Output menu.

The Graphics command invokes presentation of the results of the circuit analysis in the form of plots displayed on the screen of the computer video monitor. Selecting the Graphics command by the up and down keys and pressing the <enter> key brings up a multiple-selection menu that controls the definition of the plotted data, as illustrated in Figure B.5. Using this menu the user may select the circuit parameters and the circuit functions that he wants to observe on the monitor screen.

JADMIC2 computes the scattering parameters *S*, the admittance parameters *Y*, and the impedance parameters *Z*. All these different matrix parameters are related to the input port and to the output port of the overall circuit. Scattering parameters are default output data, whereas computation of the impedance and admittance parameters must be invoked in the input file. All of these parameters are presented in the form of plots of real and imaginary parts as functions of frequency in rectangular coordinates or in the form of radial plots on the Smith chart.

Figure B.6 presents the submenu for **S**-matrix, **Y**-matrix, and **Z**-matrix parameters. When a particular parameter from one of these menus is highlighted and selected using up and down arrow keys, pressing the <enter> key invokes presentation of these parameters in the form of a plot on the video monitor screen.

The resulting screen plot may be output on a printer using one of the many Graphics screen dump utilities designed for the PC.

The fourth selection on the Graphics submenu, called Noise parameters, is illustrated in Figure B.7.

This submenu may be used for invoking the graphics presentation of different noise parameters of the analyzed circuit. One of the most important noise param-

```
Output
 Graphics
   S-parameters
   Y-parameters
   Z-parameters
   Noise Parameters
   Circuit Functions
   Exit
```

Figure B.5 The Output Data selection menu.

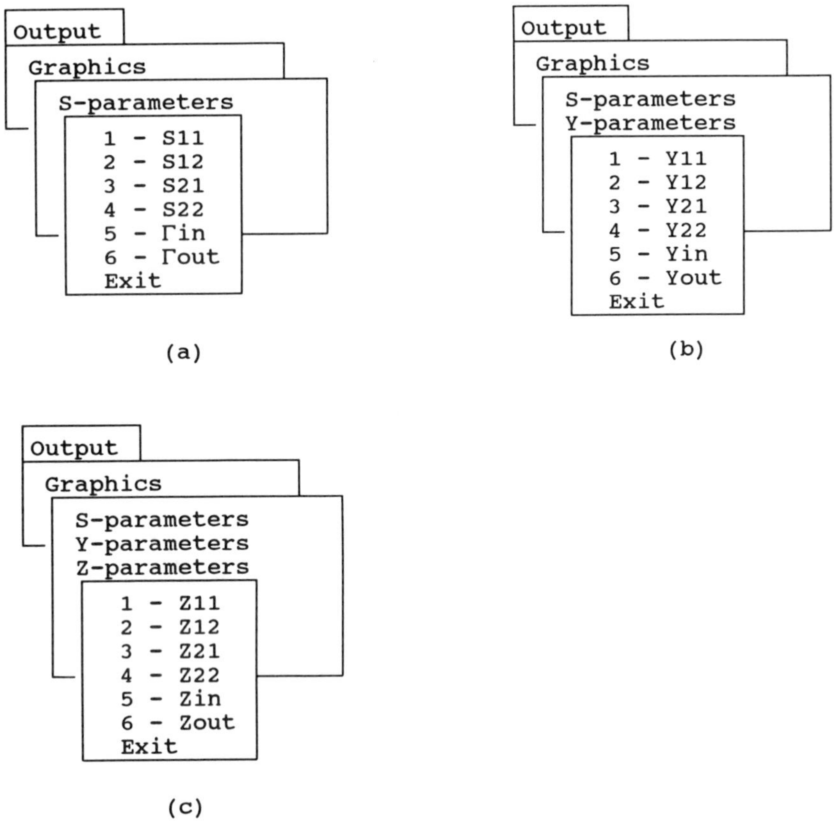

Figure B.6 Submenus for (a) **S**-matrix parameters, (b) **Y**-matrix parameters, and (c) **Z**-matrix parameters.

eters is the noise figure NF. The noise figure of the analyzed circuit corresponds to the input and to the output ports of the circuit defined in the TPL—topology block of the input file. The other noise parameters computed by JADMIC2 are the minimum noise figure, the optimum reflection coefficient corresponding to the minimum noise figure, and the noise resistance of the overall network treated as a two-port defined by its input and output ports.

The last submenu in the output Graphics menu is called Circuit functions.

Highlighting and selecting the Circuit function bar by using the up and down arrow keys and the <enter> key calls out six choices for the output parameters of the analyzed circuit. Figure B.8 presents the Circuit functions submenu.

The circuit functions that may be selected for presentation as plots in rectangular coordinates are all frequency-dependent quantities: power gain, available

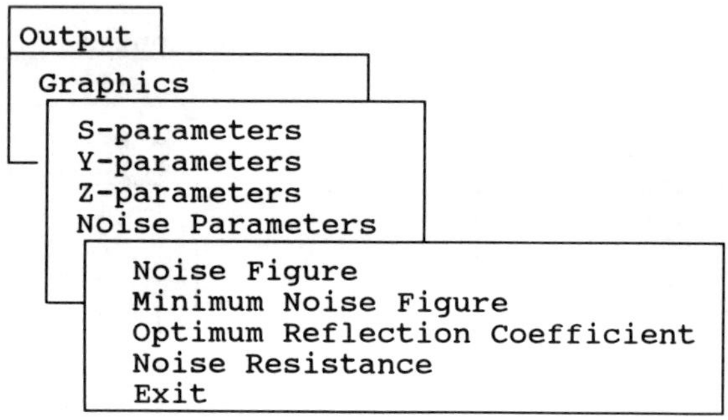

Figure B.7 Noise Parameters submenu.

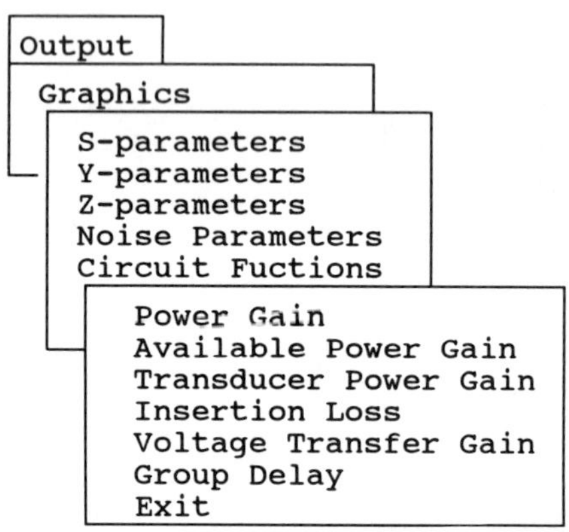

Figure B.8 Circuit Functions submenu.

power gain, transducer power gain, insertion loss, voltage transfer gain, and group delay. The above-mentioned circuit functions depend on impedances that terminate the input and output ports of the circuit.

APPENDIX C

The Input File—General Information

The input file contains the description of the circuit to be analyzed as well as instructions and specifications about the type of analysis and/or optimization, and about the type required by the user outputs.

The input file can be prepared before running the JADMIC2 program, or it can be entered and edited using the internal JADMIC editor (see Section B.1.1).

The input data file should be named according to DOS rules with the extension .CKT. When requesting the input data file during JADMIC2 execution, the user may not need to include the extension unless the file to be used has an extension other than .CKT. If the file to be used as the input file for JADMIC2 does not have an extension, the user must add . to the file name when entering that name to JADMIC2.

The main application of the internal editor is entering modifications in the input file between subsequent analyses of the circuit being designed and modified.

C.1 ORGANIZATION OF THE INPUT FILE—INPUT DATA BLOCKS

In general, the input file blocks may appear in the input file in any order. With the exception of comment lines, all other lines must be located within blocks.

The types of input data sections are summarized as:

- TITLE block (if any);
- CKT (element) block;
- TPL (topology) block;
- FREQ (frequency) block;
- OUT (output) block;
- OPT (optimization) block;
- DATA block (if any);
- LIB (library) block (if any).

CKT, TPL, FREQ, and OUT data blocks are necessary blocks of the input data file for JADMIC2. The CKT block must precede the TPL block, and the FREQ block must precede the LIB and OPT blocks. The other blocks may be arranged in any order.

Other kinds of blocks may be optionally added, but all blocks must be terminated by the END statement and a separate END statement must end the whole input data file.

The input data file is organized as in Figure C.1 and is illustrated in Figure C.2. The JADMIC2 input data file describes the transistor circuit.

TITLE block

It is recommended that the title for the problem solved using JADMIC2 be defined. By using the TITLE block, the user may define the name of the problem.

CKT block (circuit element block)

The CKT input data block starts with the mandatory key word CKT and contains statements that describe circuit elements and their parameters. Every circuit element is defined by means of an elementary model stored in the JADMIC element library or by means of measured characteristics.

TPL block

The TPL block starts from the mandatory key word TPL and describes interconnections of ports of previously defined circuit elements. In the case of microwave integrated circuit analysis and design, it also describes, for example, geometrical topology and layout, bends of microstrips, steps, and T-junctions.

FREQ block

The key word FREQ designates a block of the input data file that specifies a set of frequencies for circuit analysis.

OUT block

The key word OUT starts the input data block in which a user specifies all the required circuit functions and parameters such as overall scattering parameters of

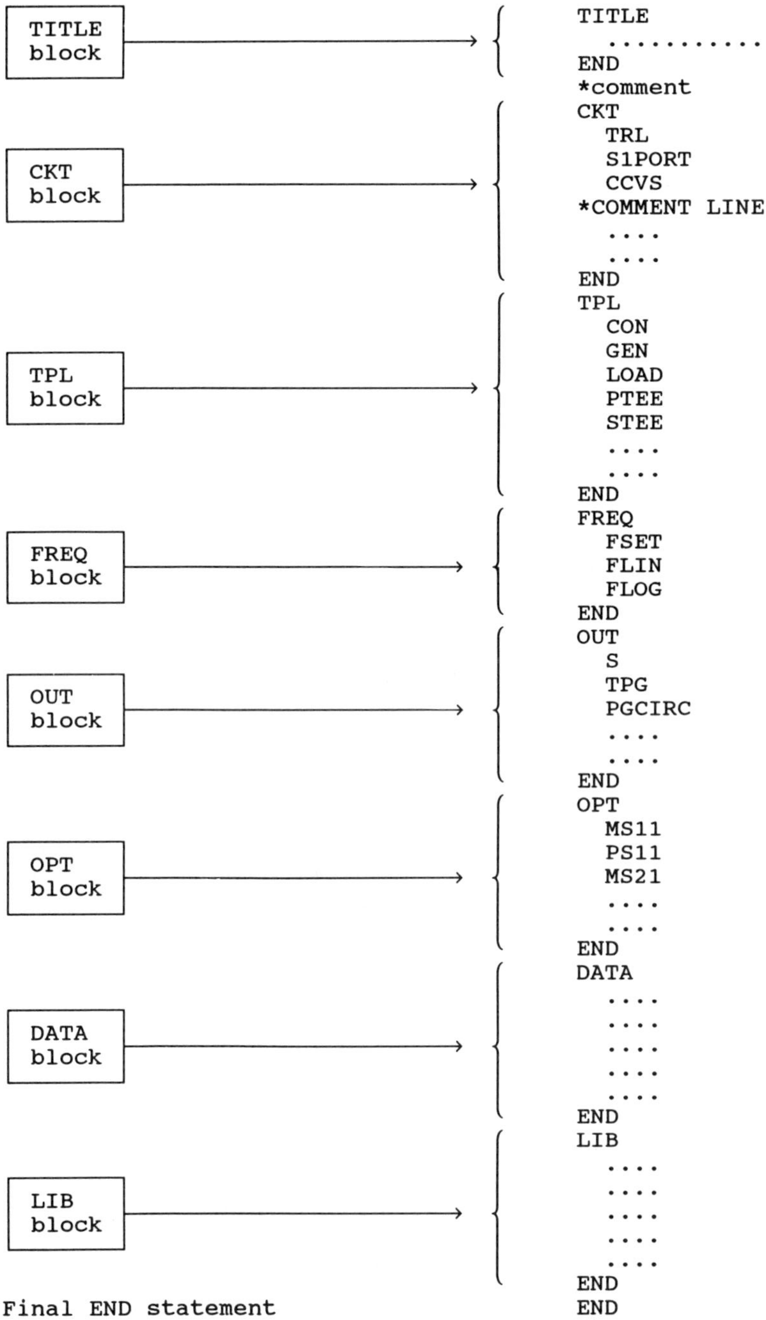

Figure C.1 Simple illustration of JADMIC2 input data file.

a circuit, group delay, noise figure, input/output reflection coefficients, constant gain circles, and constant noise figure circles.

OPT block

In the case of circuit optimization, specifications of, for example, optimization error function, limit, and goal performance are defined in the input data block that starts from the key word OPT.

DATA block

The DATA block starts with the key word DATA. It lists data common to the whole circuit and referenced by other blocks.

LIB block

The LIB block lists scattering parameters and noise parameters of frequency-dependent elements described by these parameters. Each of the LIB blocks is named by a user-defined label.

These types of input data blocks are summarized in Table C.1.

C.1.1 Comment Lines

A comment line allows a user to insert textual information into the input file for the purpose of documenting and clarifying the input data structure. Comment lines and empty records (lines) can be entered anywhere in the input data file. Comment lines must begin with a star (*) followed by a space. The comment may be entered also on the line (record). It must start from the star mark. JADMIC2 ignores the rest of such a line.

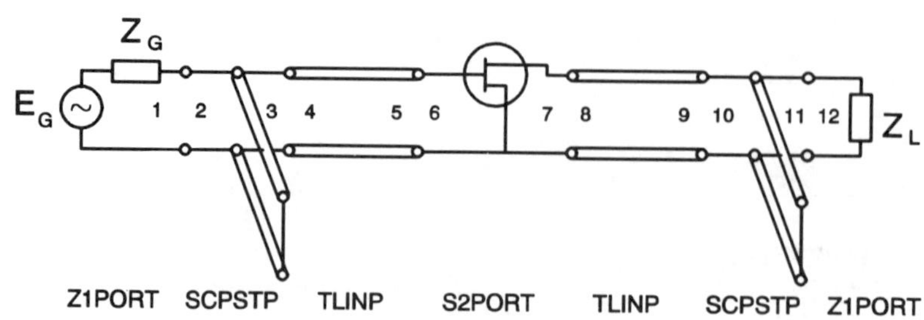

Figure C.2(a) A circuit of a transistor amplifier partitioned into JADMIC2 elements.

```
* ONE-STAGE TRANSISTOR AMPLIFIER ANALYZED AT 10 FREQUENCIES
* IN THE BAND 1. TO 2.5 GHZ

TITLE
ONE - STAGE TRANSISTOR AMPLIFIER
END

CKT
* GENERATOR AND ITS PARAMETERS
  Z1PORT 1 REZ = 50. IMZ = 0.
* SHORT CIRCUITED PARALLEL STUB
  SCPSTP 1 3 Z = 70. PL = 150 MM K = 0.5 A = 0.1 dB/M
                            F = 0.
* TRANSMISSION LINE
  TLINP 4 5 Z = 50. PL = 120 MM K = 0.55 A = 0.3 DB/M
* TRANSISTOR CALLED FET1—TWO-PORT DESCRIBED BY ITS SCATTERING
* PARAMETERS
  S2PORT 6 7 FET1
* TRANSMISSION LINE
  TLINP 8 9 Z = 50. PL = 110 MM K = 0.55 A = 0.3 DB/M
* SHORT CIRCUITED PARALLEL STUB
  SCPSTP 10 11 Z = 70. PL = 155. MM K = 0.5 A = 0.3 DB/M
* LOAD PORT PARAMETERS
  Z1PORT 12 REZ = 50. IMZ = 0.
END

TPL
  CON 1 = 2 3 = 4 5 = 6 7 = 8 9 = 10 11 = 12
  GEN = 1
  LOAD = 12
END

FREQ
  FLIN 1.0 GHZ 2.5 GHZ 10
END

OUT
  S
END

LIB
* FET1 SCATTERING PARAMETERS
FET1 ZR = 50 OHM
1.  GHZ    0.512   193.   0.075   56.   2.24   72.    0.535   340.
1.5 GHZ     .575   170.    .076   56.   1.1    67.     .530   329.
2.0 GHZ     .612   179.    .09    51.   1.4    52.5    .528   322.
2.5 GHZ     .650   168.    .09    66.   1.15   47.3    .528   332.
END
* END OF INPUT DATA FILE
END
```

Figure C.2(b) JADMIC2 input data file describing the transistor circuit.

Table C.1
Input File Blocks—Summary

Block Type	Key Word	Function	Reference
Title	TITLE	Defines a user title to be used in all outputs	Appendix C
Circuit elements	CKT	Defines circuit element types, port numbers, and element parameter values	Appendix D
Circuit topology	TPL	Defines interconnections of circuit element ports. Determines signal generator port number and load port number	Appendix E
Frequency	FREQ	Specifies frequency set for network analysis	Appendix F
Library	LIB	Lists scattering data and/or noise parameters. Groups of data are referenced by a name (label)	Appendix G
Optimization	OPT	Specifies optimization goals and optimization parameters	Appendix H
Output	OUT	Specifies performance characteristics to be computed	Appendix I
Data	DATA	Specifies data common to entire network	Appendix J

C.1.2 Continuation Lines

All lines (records) in the input file may not be more than 80 characters in length. The setup of the input data is such that each statement starts with a self-explanatory key word. If a statement requires more than 80 characters, it may be continued on the next line. The break of the lines must be made in any place where a space would normally occur. Two consecutive alphanumeric information must be separated by at least one blank.

C.1.3 Indenting

Any line of the input file can be indented any number of spaces. Indenting may be applied within and between blocks. The proper use of indenting makes the input file easier to read.

C.1.4 Numbers

The numbers in an input record are accepted as integers and as floating point numbers written without the exponent notation as well as with exponent notation. The letter E or letter D may be used to specify the exponent part of the floating point values in exponent notation.

Since blank characters are used as delimiters, they must not be embedded in a number. Element and port numbers should be introduced as integers. All other parameters should be entered as real or integer numbers.

Considerable freedom is allowed in specifying the floating point numbers in exponent notation. For example, 132., 132, .132E3, 1.32D2, and 1320d-1 represent the same number.

C.1.5 Letters

Lowercase as well as uppercase letters may be used in the key words of the input file. Blank characters must not be embedded in a key word (mnemonic) of the input data file.

C.1.6 Units

To specify numerical values of circuit parameters we must use proper units. Table C.2 lists units of different physical quantities and their mnemonics used in JADMIC2. The table indicates the default units for different quantities specified. The default unit is that which is assumed unless it is specified otherwise in the input file. Defaults are overridden by adding the unit name (mnemonic) to the numerical value. Only the nearest preceding number is affected by the added unit name.

Appendixes D to K contain a general schematic description of the input for JADMIC2. The following conventions are used in describing the individual items that may appear in a control or circuit description statement.

- Uppercase lettering indicates reserved command names; for example, the expression

  ```
  TLINP   p1   p2   Z = x1   PL = x2   K = x3   A = x4   F = x5
  ```

  means that when entering parameters of a TLINP type element np1, np2, x1, x2, x3, x4, and x5 must be replaced with numeric values as:

```
TLINP 3 4   Z = 50 OHM   PL = 1.2 INCH   K = 0.65 A = .2 DB/M   F = 5 GHz
```

Table C.2.
Units and Their Names Used in JADMIC2

Quantity	Units		Default Unit
Frequency *F*	HZ,	hertz	HZ
	kHZ,	kilohertz	
	MHZ,	megahertz	
	GHZ,	gigahertz	
Resistance *R*	OHM,	ohm	OHM
	kOHM,	kiloohm	
	MOHM,	megaohm	
Inductance *L*	H,	henry	H
	MH,	millihenry	
	UH,	microhenry	
	NH,	nanohenry	
	PH,	picohenry	
Capacitance *C*	PF,	picofarad	F
	NF,	nanofarad	
	UF,	microfarad	
	MF,	millifarad	
	F,	farad	
Physical length *PL*	UM,	micrometer ($=10^{-6}$m)	M
	MM,	millimeter	
	CM,	centimeter	
	M,	meter	
	UIN,	microinch	
	MIN,	10^{-3} inches	
	IN,	inch	
	A,	Ångström	
Electrical length *EL*	DEG,	degree	DEG
	RAD,	radian	
Conductance *G*	MHO,	1/ohm	MHO
	MMHO,	1/kiloohm	
	UMHO,	1/megaohm	
Attenuation *A*	DB/M,	decibel/meter	
	DB/CM,	decibel/centimeter	
	dB/MM,	decibel/millimeter	
	DB/IN,	decibel/inch	
	DB/MIL,	decibel/MIL	
	DB/UIN,	decibel/microinch	
	DB/A,	decibel/Ångström	
	NP/M,	neper/meter	NP/M
	NP/CM,	neper/centimeter	
	NP/MM,	neper/millimeter	
	NP/IN,	neper/inch	
	NP/MIL,	neper/milliinch	
	NP/UIN,	neper/microinch	
	NP/A,	neper/Ångström	

- Lowercase lettering indicates user-supplied information.
- Brackets [ ] enclose optional items.
- Ellipses . . . indicate that an appropriate number of the preceding entries are necessary.
- Braces { } indicate a required choice among items listed in brackets. The items are separated by vertical bars.

APPENDIX D

Specifying Network Elements—CKT Block

An input data block that defines network elements starts with the mandatory key word CKT. Every network element is defined by means of measured characteristics or by an elementary model stored in the program library.

The element types implemented in the JADMIC2 library cover an extensive repertoire of lumped and distributed elements. Some very important kinds of JADMIC2 elements are those determined in terms of their geometrical dimensions and the electrical characteristics of the materials forming them. These elements are called *microwave integrated circuit* (MIC) distributed elements and MIC lumped elements.

Each network element is defined by a statement composed of:

- A key word (mnemonic, easy to learn) describing the type of the element;
- Numbers specifying port numbers of the element;
- Key words and numbers specifying element parameters (if any).

The ports must be numbered consecutively starting from 1. Ports belonging to the same element must be numbered consecutively and according to rules stated in Section D.1.

Additionally, specifications for optimization, relating to the element, are placed on the element statement line. These special items are described in Chapter 8, which is devoted to network optimization.

Format-free statements describing the network element have the following structures.

For a one-port element:

Mnemonic determining element type	Port number	Element parameters

For a two-port element:

Mnemonic determining element type	1st port number	2nd port number	Element parameters

For a three-port element:

Mnemonic determining element type	1st port number	3rd port number	Element parameters

For a four port element:

Mnemonic determining element type	1st port number	4th port number	Element parameters

Details of entering numerical data for each JADMIC2 element are described and summarized in Section D.1.

The syntax of the CKT block is

```
CKT
  ELEMENT   PORT1   PORT2   DATA1 = x1   ............ DATAn = xN
  ELEMENT   PORT1   PORT3   DATA1 = X1   ............ DATAn = xN
  ..............................................................
  ..............................................................
  ELEMENT   PORT1   DATA1 = x1   DATAn = xN
END
```

For example,

```
CKT
  MATCH 1
  TLINP 2 3 Z= 50 OHM PL = 120 MM A= .1 dB/M K = 2.
  PRLCP 4 5 R=20 L = 5 NH C = 1.4 PF
  TRAFO 6 7 P = 9
  MATCH 8
END
```

D.1 JADMIC2 NETWORK ELEMENTS

For the sake of clarity, JADMIC2 network elements have been arranged into seven groups:

1. Passive lumped elements;
2. Measured characteristics elements;
3. Device models;
4. Distributed elements;
5. Microstrip distributed elements;
6. MIC lumped elements;
7. Universal elements.

This section contains a full description of the network elements available in the JADMIC2 program.

D.1.1 Passive Lumped Elements

CAP	Capacitor as one-port (EL30).
CAPP	Capacitor connected parallel to input and output ports (EL33).
CAPQ	Lossy capacitor as one-port (EL31).
CAPQP	Lossy capacitor connected parallel to input and output ports (EL35).
CAPQS	Lossy capacitor connected in series to input and output ports (EL34).
CAPS	Capacitor connected in series to input and output ports (EL32).
IND	Inductor as one-port (EL36).
INDP	Inductor connected parallel to input and output ports (EL39).
INDQ	Lossy inductor as one-port (EL37).
INDQP	Lossy inductor connected parallel to input and output ports (EL41).
INDQS	Lossy inductor connected in series to input and output ports (EL40).
INDS	Inductor connected in series to input and output ports (EL38).
MUC	Coupled coils (EL29).
PRLC	Parallel *RLC* circuit as one-port (EL08).
PRLCP	Parallel *RLC* circuit connected parallel to input and output ports (EL09).
PRLCS	Parallel RLC circuit connected in series to input and output ports (EL10).
RES	Resistor as one-port (EL42).
RESP	Resistor connected parallel to input and output ports (EL44).
RESS	Resistor connected in series to input and output ports (EL43).
SRLC	Series *RLC* circuit as one-port (EL05).
SRLCP	Series *RLC* circuit connected parallel to input and output ports (EL07).
SRLCS	Series *RLC* circuit connected in series to input and output ports (EL06).
TRAFO	Ideal transformer (EL19).

D.1.1.1 CAP—Capacitor

Schematic

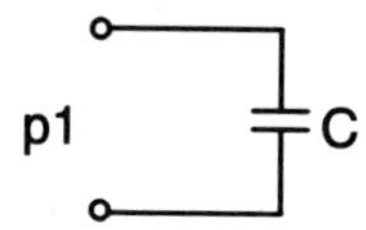

Data

p1 = Port number
C = Capacitance

Syntax

CAP p1 C = x1

Example

CAP 13 C = 1. pF

D.1.1.2 CAPP—Capacitor Parallel to Input and Output Ports

Schematic

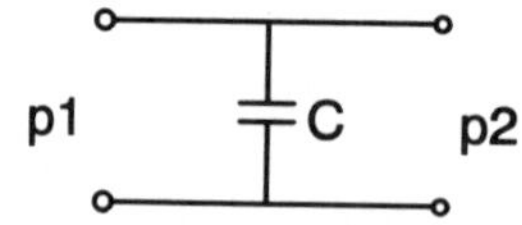

Data

p1 = Port number
p2 = Port number
C = Capacitance

Syntax

CAPP p1 p2 C = x1

Example

CAPP 13 14 C = 1. pF

D.1.1.3 CAPQ—Lossy Capacitor

Schematic

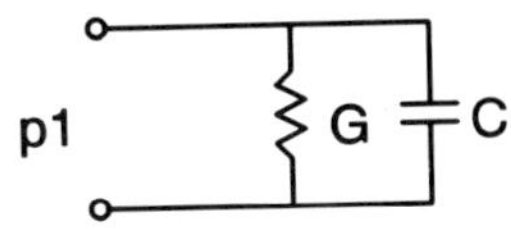

$Q = 2*\Pi * f * C/G$ where f denotes frequency

Data

p1 = Port number
C = Capacitance
Q = Quality factor
F = Frequency at which Q is given
MOD = Code number for frequency dependences of Q

Syntax

CAPQ p1 C = x1 Q = x2 F = x3 MOD = x4

Examples

CAPQ 9 C = 5 pF Q = 100 F = 1 GHz MOD = 1
CAPQ 9 C = 5 pF Q = 100 F = 1 GHz MOD = 2
CAPQ 9 C = 5 pF Q = 100 F = 1 GHz MOD = 3

Notes

1. MOD code numbers may accept values 1, 2, and 3. They correspond to the following frequency dependences of Q:

MOD = 1	$Q(f) = Q(F) * f/F$	$G(f) = G(f)$
MOD = 2	$Q(f) = Q(F) * \sqrt{f/F}$	$G(f) = G(F) * \sqrt{f/F}$
MOD = 3	$Q(f) = Q(F)$	$G(f) = G(F) * f/F$

   where f is the current frequency and F represents the value assigned to the key word F in the CAPQ statement.

2. If a value other than 1, 2, or 3 is given to MOD, JADMIC2 assumes that the capacitor is lossless (in such case CAPQ is the same as the element CAP).

D.1.1.4 CAPS—Capacitor in Series to Input and Output Ports

Schematic

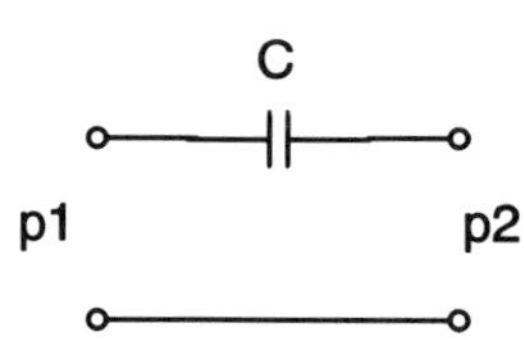

Data

p1 = Port number
p2 = Port number
C = Capacitance

Syntax

```
CAPS p1 p2 C = x1
```

Example

```
CAPS 13 14 C = 1. pF
```

D.1.1.5 CAPQP—Lossy Capacitor Parallel to Input and Output Ports

Schematic

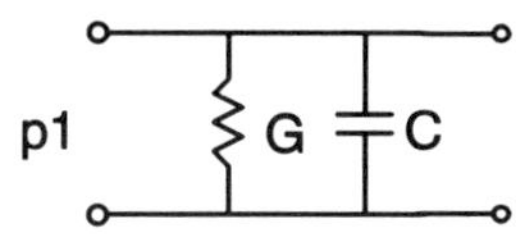

Q = 2*Π * f * C/G where f denotes frequency

Data

p1 = Port number
p2 = Port number
C = Capacitance
Q = Quality factor
F = Frequency at which Q is given
MOD = Code number for frequency dependences of Q

Syntax

CAPQP p1 p2 C = x1 Q = x2 F = x3 MOD = x4

Examples

CAPQP 9 10 C = 5 pF Q = 100 F = 1 GHz MOD = 1
CAPQP 9 10 C = 5 pF Q = 100 F = 1 GHz MOD = 2
CAPQP 9 10 C = 5 pF Q = 100 F = 1 GHz MOD = 3

Notes

1. MOD code numbers may accept values 1, 2, and 3. They correspond to the following frequency dependences of Q:

   MOD = 1 $Q(f) = Q(F) * f/F$ $G(f) = G(F)$
   MOD = 2 $Q(f) = Q(F) * \sqrt{f/F}$ $G(f) = G(F) * \sqrt{f/F}$
   MOD = 3 $Q(f) = Q(F)$ $G(f) = G(F) * f/F$

   where f is the current frequency and F represents the value assigned to the key word F in the CAPQP statement.

2. If a value other than 1, 2, or 3 is given to MOD, JADMIC2 assumes that the capacitor is lossless (in such a case CAPQP is the same as the element CAPP).

D.1.1.6 CAPQS—Lossy Capacitor in Series to Input and Output Ports

Schematic

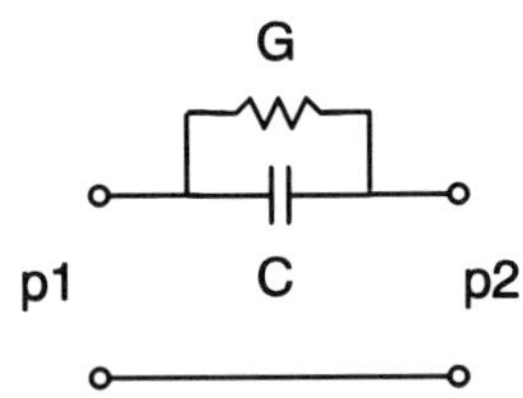

$$Q = 2 * \Pi * f * C/G \quad \text{where f denotes frequency}$$

Data

p1	= Port number
p2	= Port number
C	= Capacitance
Q	= Quality factor
F	= Frequency at which Q is given
MOD	= Code number for frequency dependence of Q

Syntax

CAPQS p1 p2 C = x1 Q = x2 F = x3 MOD = x4

Examples

CAPQS 9 10 C = 5 pF Q = 100 F = 1 GHz MOD = 1
CAPQS 9 10 C = 5 pF Q = 100 F = 1 GHz MOD = 2
CAPQS 9 10 C = 5 pF Q = 100 F = 1 GHz MOD = 3

Notes

1. MOD code numbers may accept values: 1, 2, and 3. They correspond to the following frequency dependences of Q:

MOD = 1 $Q(f) = Q(F) * f/F$ $G(f) = G(F)$
MOD = 2 $Q(f) = Q(F) * \sqrt{f/F}$ $G(f) = G(F) * \sqrt{f/F}$
MOD = 3 $Q(f) = Q(F)$ $G(f) = G(F) * f/F$

where f is the current frequency and F represents the value assigned to the key word F in CAPQS statement.

2. If a value other than 1, 2, or 3 is given to MOD, JADMIC2 assumes that the capacitor is lossless (in such a case, CAPQS is the same as the element CAPS).

D.1.1.7 IND—Inductor

Schematic

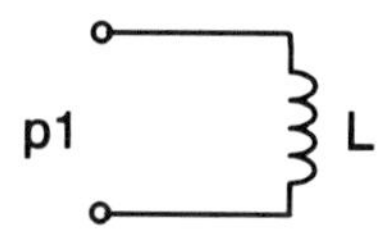

Data

p1 = Port number
L = Inductance

Syntax

IND p1 L = x1

Example

IND 13 L = 1. nH

D.1.1.8 INDP—Inductor Parallel to Input and Output Ports

Schematic

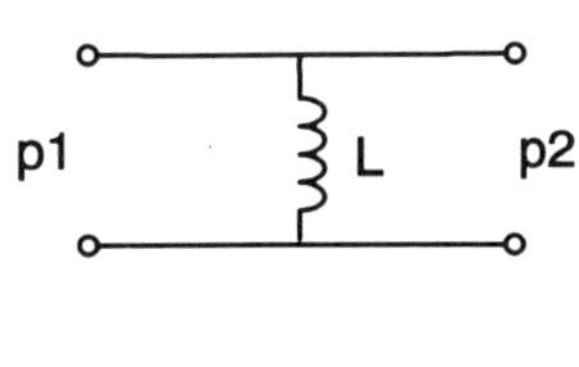

Data

p1 = Port number

p2 = Port number
L = Inductance

Syntax

INDP p1 p2 L = x1

Example

INDP 17 18 L = 1. nH

D.1.1.9 INDQ—Lossy Inductor

Schematic

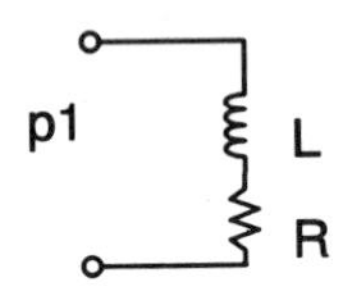

Q = 2*Π * f * L/R where f denotes frequency

Data

p1 = Port number
L = Inductance
Q = Quality factor
F = Frequency at which Q is given
MOD = Code defining frequency dependences of Q

Syntax

INDQ p1 L = x1 Q = x2 F = x3 MOD = x4

Example

INDQ 17 L = 1. nh Q = 50 F = 1 MHz MOD = 1
INDQ 17 L = 1. nH Q = 50 F = 1 MHz MOD = 2
INDQ 17 L = 1. nH Q = 50 F = 1 MHz MOD = 3

Notes

1. MOD code numbers may accept values 1, 2, and 3. They correspond to the following frequency dependences of Q:

   MOD = 1 Q(f) = Q(F) * f/F R(f) = R(F)
   MOD = 2 Q(f) = Q(F) *$\sqrt{f/F}$ R(f) = R(F) *$\sqrt{f/F}$
   MOD = 3 Q(f) = Q(F) R(f) = R(F) * f/F

   where f is the current frequency and F represents the value assigned to the key word F in the INDQ statement.

2. If a value other than 1, 2, or 3 is given to MOD, JADMIC2 assumes that the capacitor is lossless (in such case INDQ is the same as the element IND).

D.1.1.10 INDQP—Lossy Inductor Parallel to Input and Output Ports

Schematic

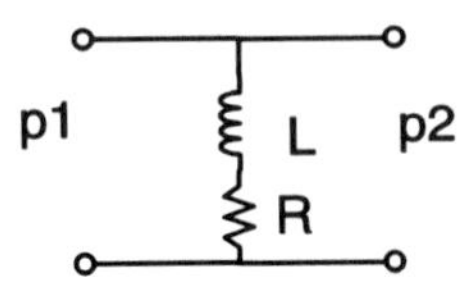

Data

p1 = Port number
p2 = Port number
L = Inductance
Q = Quality factor
F = Frequency at which Q is given
MOD = Code defining frequency dependences of Q

Syntax

INDQP p1 p2 L = x1 Q = x2 F = x3 MOD = x4

Example

INDQP 17 18 L = 1. nH Q = 50 F = 1 MHz MOD = 1
INDQP 17 18 L = 1. nH Q = 50 F = 1 MHz MOD = 2
INDQP 17 18 L = 1. nH Q = 50 F = 1 MHz MOD = 3

Notes

1. MOD code numbers may accept values 1, 2, and 3. They correspond to the following frequency dependences of Q:

   MOD = 1 $Q(f) = Q(F) * f/F$ $R(f) = R(F)$
   MOD = 2 $Q(f) = Q(F) * \sqrt{f/F}$ $R(f) = R(F) * \sqrt{f/F}$
   MOD = 3 $Q(f) = Q(F)$ $R(f) = R(F) * f/F$

   where f is the current frequency and F represents the value assigned to the key word F in the INDQP statement.

2. If a value other than 1, 2, or 3 is given the MOD, JADMIC2 assumes that the capacitor is lossless (in such case INDQP is the same as the element INDP).

D.1.1.11 INDQS—Lossy Inductor in Series to Input and Output Ports

Schematic

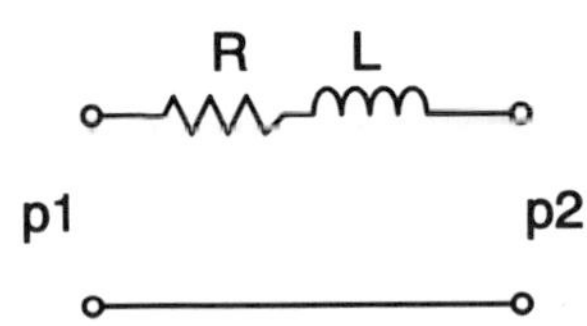

Data

p1 = Port number
p2 = Port number
L = Inductance
Q = Quality factor
F = Frequency at which Q is given
MOD = Code defining frequency dependences of Q

Syntax

INDQS p1 p2 L = x1 Q = x2 F = x3 MOD = x4

Example

INDQS 17 18 L = 1. nH Q = 50 F = 1 MHz MOD = 1
INDQS 17 18 L = 1. nH Q = 50 F = 1 MHz MOD = 2
INDQS 17 18 L = 1. nH Q = 50 F = 1 MHz MOD = 3

Notes

1. MOD code numbers may accept values 1, 2, and 3. They correspond to the following frequency dependences of Q:

   MOD = 1 $Q(f) = Q(F) * f/F$ $R(f) = R(F)$
   MOD = 2 $Q(f) = Q(F) * \sqrt{f/F}$ $R(f) = R(F) * \sqrt{f/F}$
   MOD = 3 $Q(f) = Q(F)$ $R(f) = R(F) * f/F$

   where f is the current frequency and F represents the value assigned to the key word F in the INDQS statement.

2. If a value other than 1, 2, or 3 is given to MOD, JADMIC2 assumes that the capacitor is lossless (in such a case, INDQS is the same as the element INDS).

D.1.1.12 INDS—Inductor in Series to Input and Output Ports

Schematic

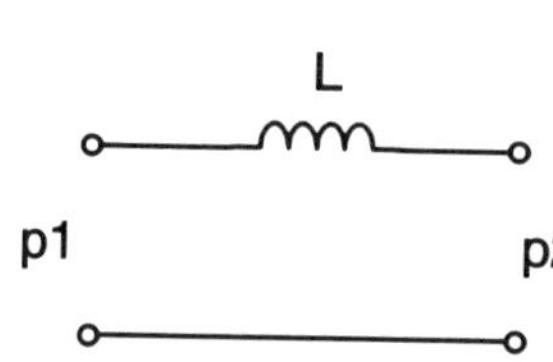

Data

p1 = Port number
p2 = Port number
L = Inductance

Syntax

INDS p1 p2 L = x1

Example

INDS 17 18 L = 1. nH

D.1.1.13 MUC—Coupled Coils

Schematic

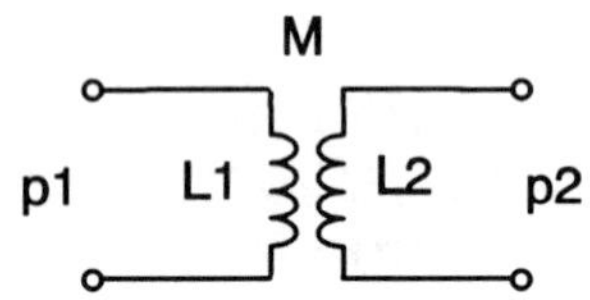

Parameters

p1 = Port number
p2 = Port number
L1 = Self-inductance of coil 1
L2 = Self-inductance of coil 2
M = Mutual inductance

Syntax

MUC p1 p2 L1 = x1 L2 = x2 M = x3

Example

MUC 3 4 L1 = 0.1 nH L2 = 5. nH M = 0.05 nH

Note

A negative value of M reverses the relative polarity shown in the schematic.

D.1.1.14 RES—Resistor

Schematic

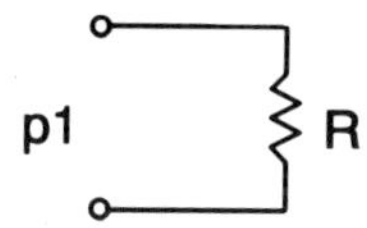

Data

p1 = Port number
R = Resistance

Syntax

RES p1 R = x1

Example

RES 21 R = 25 Ohm

D.1.1.15 RESP—Resistor Parallel to Input and Output Ports

Schematic

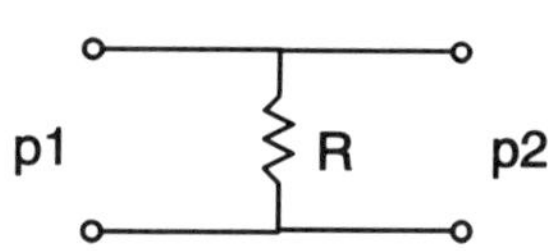

Data

p1 = Port number
p2 = Port number
R = Resistance

Syntax

RESP p1 p2 R = x1

Example

RESP 21 22 R = 25 Ohm

D.1.1.16 RESS—Resistor in Series to Input and Output Ports

Schematic

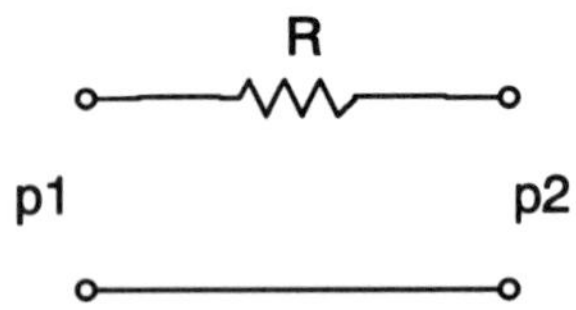

Data

p1 = Port number
p2 = Port number
R = Resistance

Syntax

RESS p1 p2 R = x1

Example

RESS 21 22 R = 25 Ohm

D.1.1.17 PRLC—Parallel RLC Circuit as One-Port

Schematic

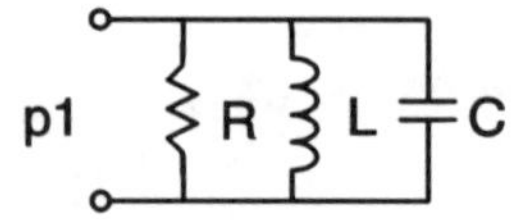

Parameters

p1 = Port number
R = Resistance
L = Inductance
C = Capacitance

Syntax

PRLC p1 [R = x1] [L = x2] [C = x3]

Example

PRLC 3 R = 0.1 L = 5. nH

Note

If one of the *RLC* parameters is not defined in the PRLC statement, it is assumed that the corresponding R, L, or C element is open circuited. In the above example the PRLC type element is composed of a resistance R = 0.1 Ω connected in parallel to the inductance L = 5. nH. A capacitance C does not exist in this example element.

D.1.1.18 PRLCP—Parallel RLC Circuit as Two-Port

Schematic

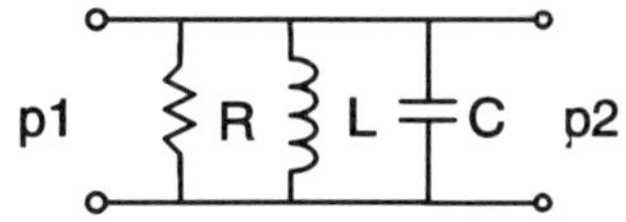

Parameters

p1 = Port number
p2 = Port number
R = Resistance
L = Inductance

C = Capacitance

Syntax

PRLCP p1 p2 [R = x1] [L = x2] [C = x3]

Example

PRLCP 3 4 R = 0.1 L = 5. nH C = 2.6 pF

Note

If one of the *RLC* parameters is not defined in the PRLCP statement, it is assumed that the corresponding R, L, or C element is open circuited.

D.1.1.19 PRLCS—Parallel RLC Circuit as Two-Port

Schematic

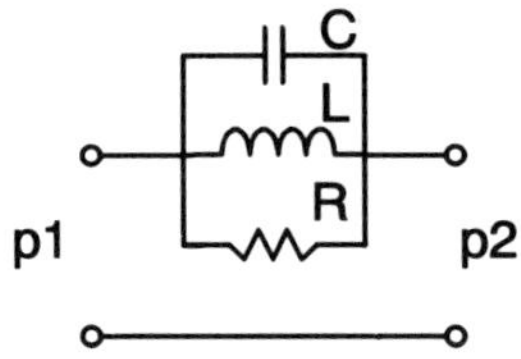

Parameters

p1 = Port number
p2 = Port number
R = Resistance
L = Inductance
C = Capacitance

Syntax

PRLCS p1 p2 [R = x1] [L = x2] [C = x3]

Example

PRLCS 3 4 R = 0.1 L = 5. nH C = 2.6 pF

Note

If one of the *RLC* parameters is not defined in the PRLCS statement, it is assumed that the corresponding R, L, or C element is open circuited.

D.1.1.20 SRLC—Series RLC Element as One-Port

Schematic

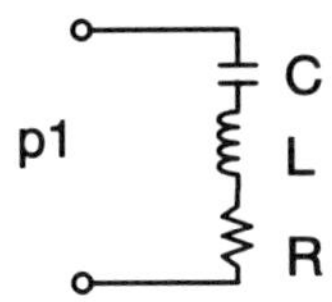

Parameters

p1 = Port number
R = Resistance
L = Inductance
C = Capacitance

Syntax

SRLC p1 [R = x1] [L = x2] [C = x3]

Example

SRLC 3 R = 0.1 L = 5. nH

Note

If one of the *RLC* parameters is not defined in the SRLC statement, it is assumed that corresponding R, L, or C element is short circuited. In the above example, the

SRLC type element is composed of a resistance R = 0.1 Ω connected in series to the inductance L = 5. nH. A capacitance C is short circuited.

D.1.1.21 SRLCP—Series RLC Circuit as Two-Port

Schematic

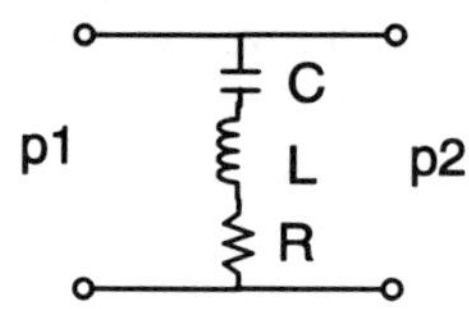

Parameters

p1 = Port number
p2 = Port number
R = Resistance
L = Inductance
C = Capacitance

Syntax

SRLCP p1 p2 [R = x1] [L = x2] [C = x3]

Example

SRLCP 13 14 R = 0.1 L = 5. nH C = 3.5 pF

Note

If one of the *RLC* parameters is not defined in the SRLCS statement, it is assumed that corresponding R, L, or C element is short circuited.

D.1.1.22 SRLCS—Series RLC Circuit as Two-Port

Schematic

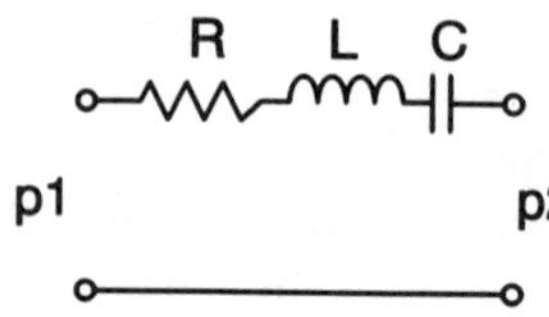

Parameters

p1 = Port number
p2 = Port number
R = Resistance
L = Inductance
C = Capacitance

Syntax

```
SRLCS p1 p2 [R = x1] [L = x2] [C = x3]
```

Example

```
SRLCS 3 4 R = 0.1 L = 5. nH C = 3.5 pF
```

Note

If one of the *RLC* parameters is not defined in the SRLCS statement, it is assumed that the corresponding R, L, or C element is short circuited.

D.1.1.23 TRAFO—Ideal Transformer

Schematic

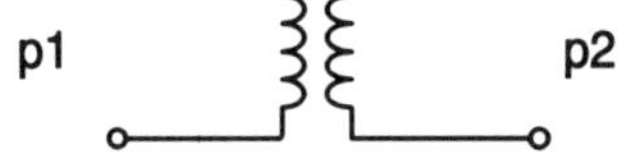

Data

p1 = Port number
p1 = Port number
p = Transformation ratio

Example

TRAFO 15 16 P = 2.5

D.1.2 Measured Characteristics Elements: One-Ports and Two-Ports Described by Their Measured Scattering Parameters

Measured characteristic elements in JADMIC2 are multiports described by the impedances, admittances, or the scattering parameters defined at discrete frequencies.

S1PORT One-port element described by its frequency-dependent scattering parameter (EL27).
S2PORT Two-port element described by its frequency-dependent 2 × 2 scattering matrix parameters (EL25).
Y1PORT One-port element described by its admittance (EL03).
Y2PORT Two-port element described by the admittance connected parallel to the input and to the output ports (EL04).
Z1PORT One-port element described by its impedance (EL01).
Z2PORT Two-port element described by the impedance connected in series to the input and to the output ports (EL02).

Measured or calculated scattering matrix multiport data are contained in the LIB block of input data. The data contained in the LIB block are referenced by labels.

For all measured characteristic elements, frequency points must be given in ascending sequence. However, a single frequency point is allowed, too. In such a case, element data are then treated as constant through the frequency band of analysis.

D.1.2.1 S1PORT—One-Port Described by S Parameters

Schematic

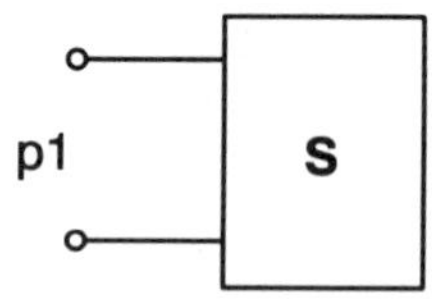

Data

p1 = Port number
label

Syntax

S1PORT p1 label

Example

S1PORT 23 DIODE1

Note

Scattering parameter data are contained in the LIB block. The data contained in the LIB block are referenced individually by a name given by the user, called the label. In the example, DIODE1 is the label that refers to the data contained in the LIB block. For LIB block formatting instructions, refer to Section G.1.1.

D.1.2.2 S2PORT—Two-Port Described by S Parameters

Schematic

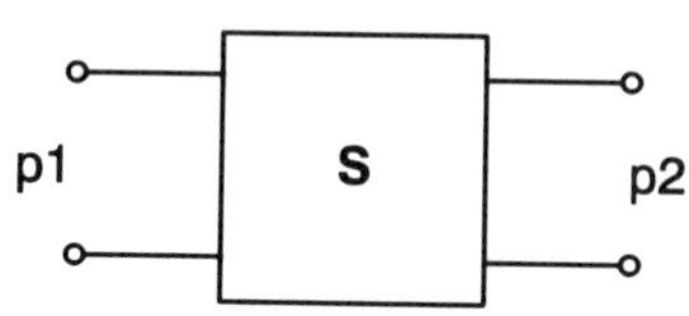

Data

p1 = Port number
p2 = Port number
label

Syntax

S2PORT p1 p2 label

Example

S2PORT 23 24 FET3

Note

Scattering parameter data are contained in the LIB block. The data contained in the LIB block are referenced individually by a name given by the user, called the label. In the example, FET3 is the label that refers to the data contained in the LIB block. For LIB block formatting instructions, refer to Section G.1.2.

D.1.2.3 Y1PORT—One-Port Defined by Its Admittance

Schematic

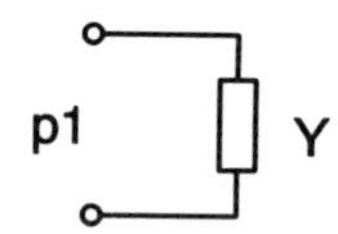

Data

p1 = Port number
REY = Real part of the admittance Y
IMY = Imaginary part of the admittance Y

Syntax

Y1PORT p1 REY = x1 IMY = x2

Example

Y1PORT 1 REY = 50 mho IMY = 230.

D.1.2.4 Y2PORT—Two-Port Defined by Its Admittance

Schematic

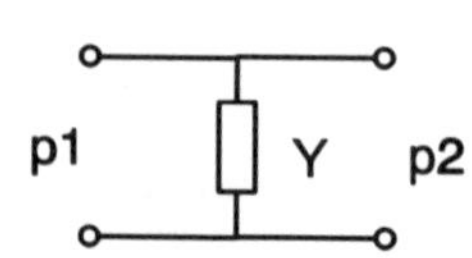

Data

p1 = Port number

p2 = Port number
REY = Real part of the admittance Y
IMZ = Imaginary part of the admittance Y

Syntax

Y2PORT p1 p2 REY = x1 IMY = x2

Example

Y2PORT 12 13 REY = 50 mho IMY = 0.125 mho

D.1.2.5 Z1PORT—One-Port Defined by Its Impedance

Schematic

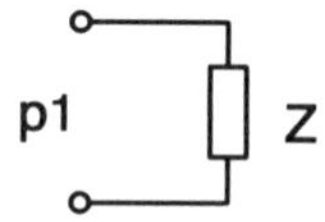

Data

p1 = Port number
REZ = Real part of the impedance Z
IMZ = Imaginary part of the impedance Z

Syntax

Z1PORT p1 REZ = x1 IMZ = x2

Example

Z1PORT 1 REZ = 50 OHM IMZ = 0

D.1.2.6 Z2PORT—Two-Port Defined by Its Impedance

Schematic

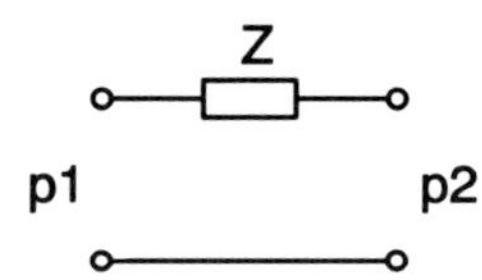

Data

p1 = Port number
p2 = Port number
REZ = Real part of the impedance Z
IMZ = Imaginary part of the impedance Z

Syntax

Z2PORT p1 p2 REZ = x1 IMZ = x2

Example

Z2PORT 1 2 REZ = 50 OHM IMZ = 220

D.1.3 Device Models

BIP Bipolar transistor (EL45).
CCCS Current-controlled current source (EL23).
CCVS Current-controlled voltage source (EL22).
PINP PIN diode connected parallel to the input and to the output ports (EL53).
PINS PIN diode connected in series to the input and to the output ports (EL52).
VCCS Voltage-controlled current source (EL21).
VCVS Voltage-controlled voltage source (EL20).

D.1.3.1 BIP—Bipolar Transistor

Schematic

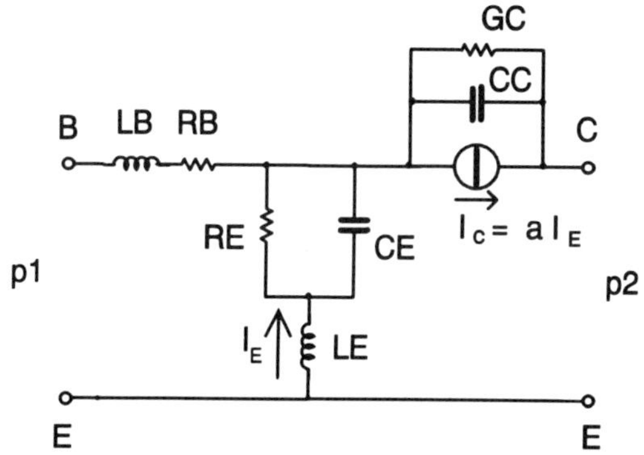

Data

p1 = Port number
p2 = Port number
A = Magnitude of current gain at dc
T = Time delay associated with current gain
F = 3-dB frequency for current gain
CC = Collector capacitance
GC = Collector conductance
RB = Base resistance
LB = Base inductance
CE = Emitter conductance
RE = Emitter resistance
LE = Emitter inductance

Syntax

```
BIP p1 p2 A = x1 T = x2 F = x3 CC = x4 GC = x5 RB = x6
LB = x7 CE = x8 RE = x9 LE = x10
```

Example

```
BIP p1 p2 A = 0.98 T = 11 us F = 1.4 GHz CC = 0.04 pF
GC = 1E-4 MHO RB = 5 LB = 0.2 nH CE = 4. pF RE = 1 LE = 0.2 nH
```

Notes

1. The current gain dependence on frequency is given as

   A(f) = A(0) * EXP(− j*2*Π*T) / (1 + j f/F)

   where f is the current frequency.
2. Assigning F = 0 results as A(f) = A(0) = const.

D.1.3.2 CCCS—Current-Controlled Current Source

Schematic

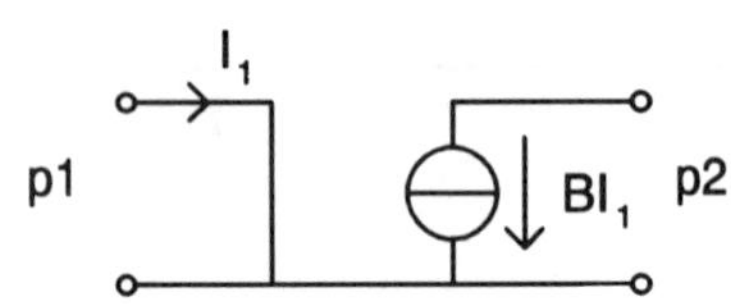

Data

p1 = Port number
p2 = Port number
B = Controlling coefficient

Syntax

CCCS p1 p2 B = x1

Example

CCCS 7 8 B = 4

D.1.3.3 CCVS—Current-Controlled Voltage Source

Schematic

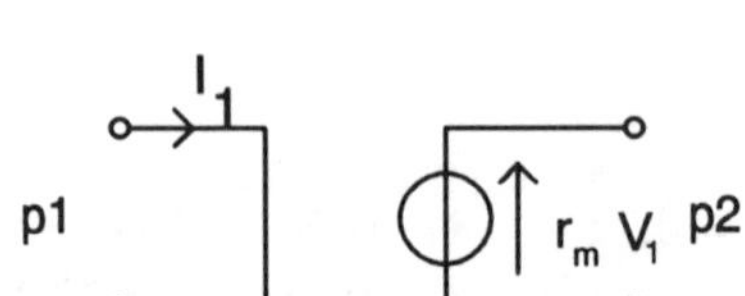

Data

p1 = Port number
p2 = Port number
RM = Transresistance

Syntax

CCVS p1 p2 RM = x1

Example

CCVS 7 8 RM = 1.0 Kohm

D.1.3.4 PINP—PIN Diode Parallel to Input and Output Port

Schematic

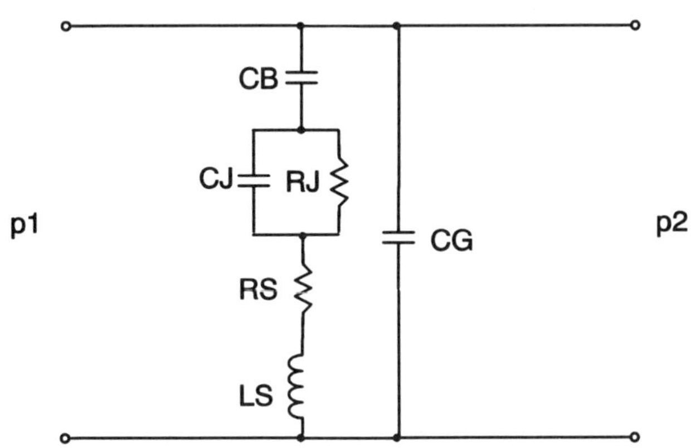

Data

p1 = Port number
p2 = Port number
CJ = Junction capacitance

RJ = Junction resistance
RS = Diode series resistance
LS = Diode series inductance
CG = Package capacitance
CB = Diode series capacitance

Syntax

PINP p1 p2 CJ = x1 RJ = x2 RS = x3 LS = x4 CG = x5 CB = x6

Example

PINP 11 12 CJ = 0.09 pF RJ = .01 RS = .6 LS = .3 nH CG = 0.2 CB = 0.5

D.1.3.5 PINS—PIN Diode in Series to Input and Output Ports

Schematic

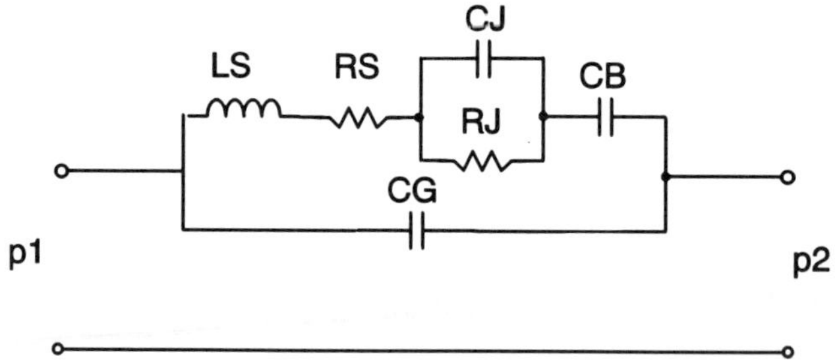

Data

p1 = Port number
p2 = Port number
CJ = Junction capacitance
RJ = Junction resistance
RS = Diode series resistance
LS = Diode series inductance
CG = Package capacitance
CB = Diode series capacitance

Syntax

PINS p1 p2 CJ = x1 RJ = x2 RS = x3 LS = x4 CG = x5 CB = x6

Example

PINS 11 12 CJ = 0.09 pF RJ = .01 RS = .6 LS = .3 nH CG = 0.2 CB = 0.5

D.1.3.6 VCVS—Voltage-Controlled Voltage Source

Schematic

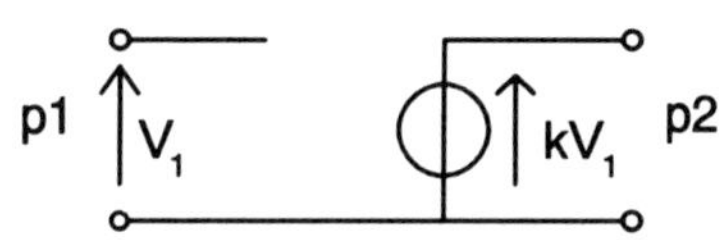

Data

p1 = Port number
p2 = Port number
k = Controlling coefficient

Syntax

VCVS p1 p2 K = x1

Example

VCVS 7 8 K = 4

D.1.3.7 VCCS—Voltage-Controlled Current Source

Schematic

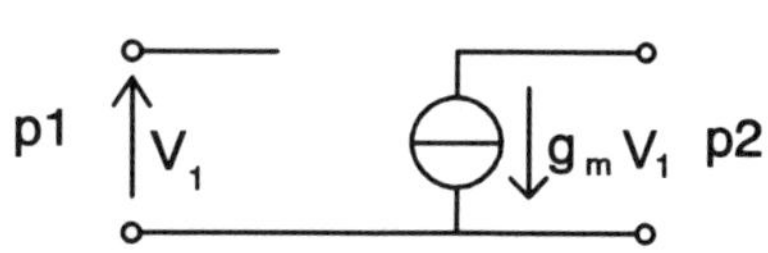

Data

p1 = Port number
p2 = Port number
GM = Transconductane

Syntax

VCCS p1 p2 GM = x1

Example

VCCS 7 8 GM = .4 MMHO

D.1.4 Distributed Elements

CTLIN	Ideal coupled transmission lines (EL61).
CTLINP	Coupled transmission lines physical model (EL18).
OCLIN	Ideal open-circuited transmission line (EL60).
OCLINP	Open-circuited transmission line physical model (EL17).
OCPST	Ideal open-circuited parallel stub (EL56).
OCPSTP	Open-circuited parallel stub physical model (EL13).
OCSST	Ideal open-circuited series stub (EL58).
OCSSTP	Open-circuited series stub physical model (EL15).
RCLIN	*RC* distributed line (EL62).
SCLIN	Ideal short-circuited transmission line (EL59).
SCLINP	Short-circuited transmission line physical model (EL16).
SCPST	Ideal short-circuited parallel stub (EL55).
SCPSTP	Short-circuited parallel stub physical model (EL12).
SCSST	Ideal short-circuited series stub (EL57).
SCSSTP	Short-circuited series stub physical model (EL14).
TLIN	Ideal transmission line (EL54).
TLINP	Transmission line physical model (EL11).

D.1.4.1 CTLIN—Coupled Transmission Lines

Schematic

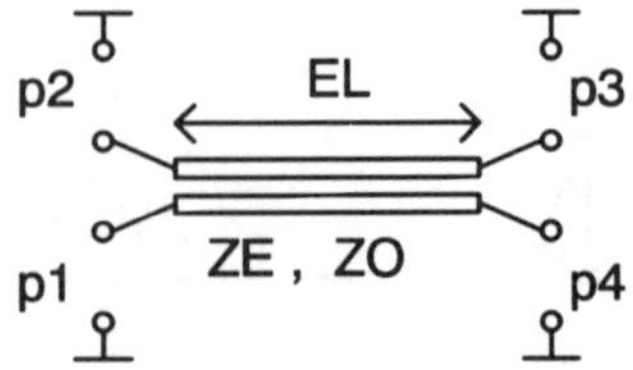

Data

p1 = Port number

p4 = Port number
ZE = Even-mode characteristic impedance
ZO = Odd-mode characteristic impedance
EL = Electrical length in radians or in degrees
F = Frequency at which electrical length is EL

Syntax

CTLIN p1 p4 ZE = x1 ZO = x2 EL = x3 F = x4

Example

CTLIN 11 14 ZE = 50ohm ZO = 10 ohm EL = 120 deg F = 2.3 GHz

Notes

1. CTLIN are ideal coupled transmission lines.
2. Even- and odd-mode phase velocities are assumed to be equal.

D.1.4.2 CTLINP—Coupled Transmission Lines Physical Model

Schematic

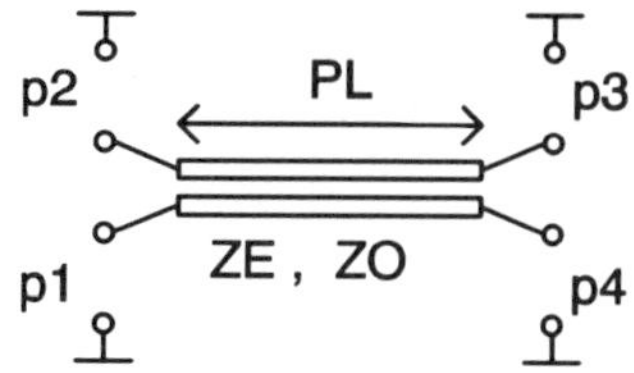

Data

p1 = Port number
p4 = Port number
ZE = Even-mode characteristic impedance
ZO = Odd-mode characteristic impedance
L = Physical length
KE = Even-mode effective dielectric constant
KO = Odd-mode effective dielectric constant
AE = Even-mode attenuation constant

AO = Odd-mode attenuation constant

Syntax

CTLINP p1 p4 ZE = x1 ZO = x2 L = x3 KE = x4 KO = x5 AE = x6 AO = x7

Example

CTLINP 11 14 ZE = 50ohm ZO = 10 ohm L = 120 mm KE = 2.3 KO = 1.91 AE = .2 dB/m AO = .2 dB/M

Note

1. Attenuation A is constant with frequency if F is not defined in the CTLINP statement or if F is zero. Otherwise, A is computed as frequency dependent as
   A(f) = A for F = 0 or undefined
   $A(f) = A * \sqrt{f/F}$ for $F \neq 0$

D.1.4.3 OCLIN—Ideal Open-Circuited Transmission Line

Schematic

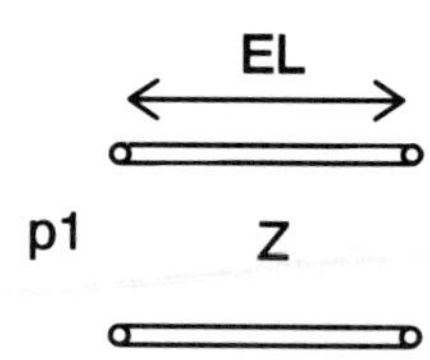

Data

p1 = Port number
Z = Characteristic impedance
EL = Electrical length in radians or in angle units
F = Frequency at which electrical length is EL

Syntax

OCLIN p1 Z = x1 EL = x2 F = x3

Example

OCLIN 11 Z = 50 EL = 90 deg F = 2.3 GHz

Note

OCLIN is an ideal open-circuited transmission line.

D.1.4.4 OCLINP—Open-Circuited Transmission Line Physical Model

Schematic

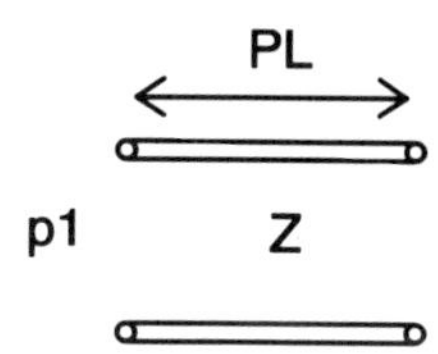

Data

p1 = Port number
Z = Characteristic impedance
PL = Physical length
K = Effective dielectric constant
A = Attenuation constant
F = Frequency for attenuation scaling

Syntax

OCLINP p1 Z = x1 PL = x2 K = x3 A = x4 F = x5

Example

OCLINP 11 Z = 50 PL = 12 mm K = 2.3 A = .3 dB/m F = 1.GHz

Note

Attenuation A is constant with frequency if F is not defined in the OCLINP statement or if F is zero. Otherwise, A is computed as frequency dependent as

$A(f) = A$ for $F = 0$ or undefined

$A(f) = A * \sqrt{f/F}$ for $F \neq 0$

D.1.4.5 OCPST—Ideal Open-Circuited Parallel Stub

Schematic

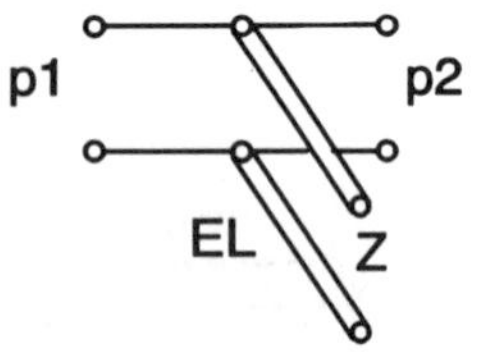

Data

p1 = Port number
p2 = Port number
Z = Characteristic impedance
EL = Electrical length
F = Frequency at which electrical length is EL

Syntax

OCPTS p1 p2 Z = x1 EL = x2 F = x3

Example

OCPTS 11 12 Z = 50 EL = 90 deg F = 2.3 GHz

Note

OPTS is an ideal open-circuited parallel stub.

D.1.4.6 OCPSTP—Open-Circuited Parallel Stub Physical Model

Schematic

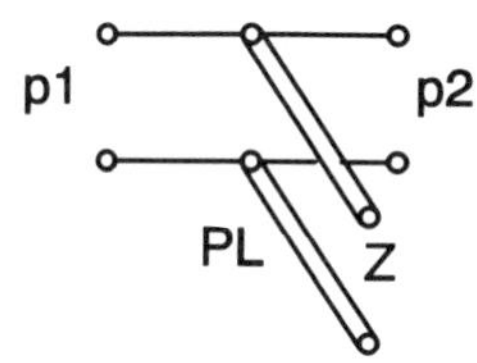

Data

p1 = Port number
p2 = Port number
Z = Characteristic impedance
PL = Physical length
K = Effective dielectric constant
A = Attenuation constant
F = Frequency for attenuation scaling

Syntax

OCPSTP p1 p2 Z = x1 PL = x2 K = x3 A = x4 F = x5

Example

OCPSTP 11 12 Z = 50 PL = 12 mm K = 2.3 A = .3 dB/m F = 1 GHz

Note

Attenuation A is constant with frequency if F is not defined in the OCPSTP statement or if F is zero. Otherwise, A is computed as frequency dependent as

$A(f) = A$ for $F = 0$ or undefined

$A(f) = A * \sqrt{f/F}$ for $F \neq 0$

D.1.4.7 OCSST—Ideal Open-Circuited Series Stub

Schematic

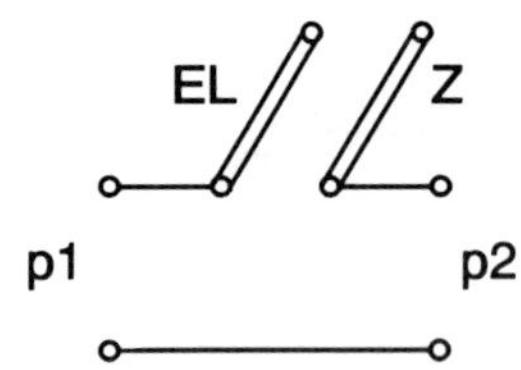

Data

p1 = Port number
p2 = Port number
Z = Characteristic impedance
EL = Electrical length in radians or in angle units
F = Frequency at which electrical length is EL

Syntax

OCSST p1 p2 Z = x1 EL = x2 F = x3

Example

OCSST 11 12 Z = 50 EL = 45 deg F = 2.3 GHz

Note

OCSST is an ideal open-circuited series stub.

D.1.4.8 OCSSTP—Open-Circuited Series Stub Physical Model

Schematic

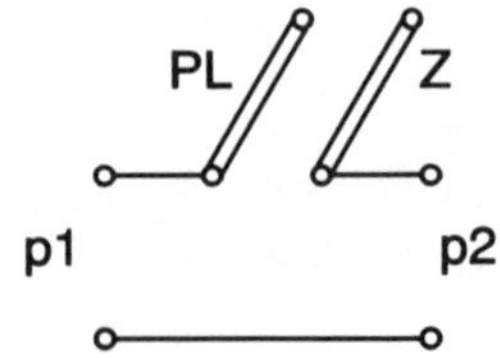

Data

p1 = Port number
p2 = Port number
Z = Characteristic impedance
PL = Physical length
K = Effective dielectric constant
A = Attenuation constant
F = Frequency for attenuation scaling

Syntax

OCSSTP p1 p2 Z = x1 PL = x2 K = x3 A = x4 F = x5

Example

OCSSTP 11 12 Z = 50 PL = 12 mm K = 2.3 A = .3 dB/m F = 1 GHz

Note

Attenuation A is constant with frequency if F is not defined in the OCSSTP statement or if F is zero. Otherwise, A is computed as frequency dependent as

$A(f) = A$ for $F = 0$ or undefined
$A(f) = A * \sqrt{f/F}$ for $F \neq 0$

D.1.4.9 RCLIN—Distributed RC Line

Schematic

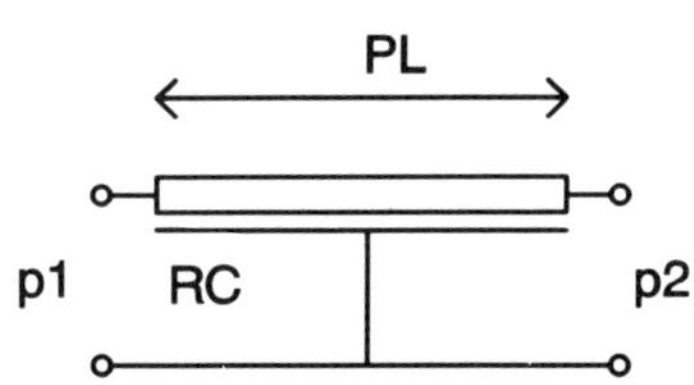

Data

p1 = Port number

p2 = Port number
R = Series resistance per unit length
C = Shunt capacitance per unit length
PL = Physical length

Syntax

RCLIN p1 p2 R = x1 C = x2 PL = x3

Example

RCLIN 8 9 R = 0.06 ohm C = 3E-3 F PL = 20 cm

D.1.4.10 SCLIN—Ideal Short-Circuited Transmission Line

Schematic

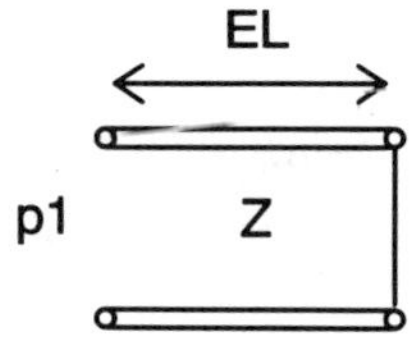

Data

p1 = Port number
Z = Characteristic impedance
EL = Electrical length in radians or in angle units
F = Frequency at which electrical length is EL

Syntax

SCLIN p1 Z = x1 EL = x2 F = x3

Example

SCLIN 11 Z = 50 EL = 90 deg F = 2.3 GHz

Note

SCLIN is an ideal short-circuited transmission line.

D.1.4.11 SCLINP—Short-Circuited Transmission Line Physical Model

Schematic

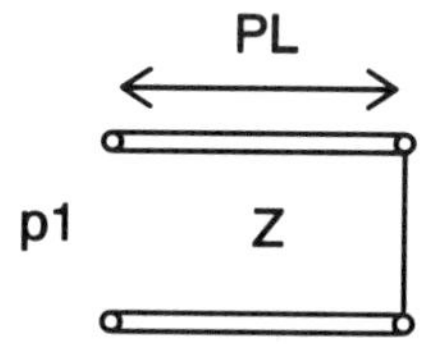

Data

p1 = Port number
Z = Characteristic impedance
PL = Physical length
K = Effective dielectric constant
A = Attenuation constant
F = Frequency for attenuation scaling

Syntax

```
SCLINP p1 Z = x1 PL = x2 K = x3 A = x4 F = x5
```

Example

```
SCLINP 11 Z = 50 PL = 12 mm K = 2.3 A = .3 dB/m F = 1 GHz
```

Note

Attenuation A is constant with frequency if F is not defined in the SCLINP statement or if F is zero. Otherwise, A is computed as frequency dependent as

$A(f) = A$ for $F = 0$ or undefined

$A(f) = A * \sqrt{f/F}$ for $F \neq 0$

D.1.4.12 SCPST—Ideal Short-Circuited Parallel Stub

Schematic

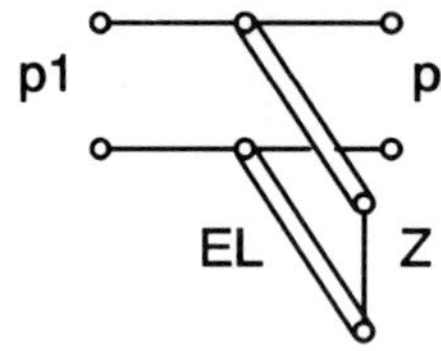

Data

p1 = Port number
p2 = Port number
Z = Characteristic impedance
EL = Electrical length in radians or in angle units
F = Frequency at which electrical length is EL

Syntax

SCPST p1 p2 Z = x1 EL = x2 F = x3

Example

SCPST 11 12 Z = 50 EL = 45 deg F = 2.3 GHz

Note

SCPST is an ideal short-circuited parallel stub.

D.1.4.13 SCPSTP—Short-Circuited Parallel Stub Physical Model

Schematic

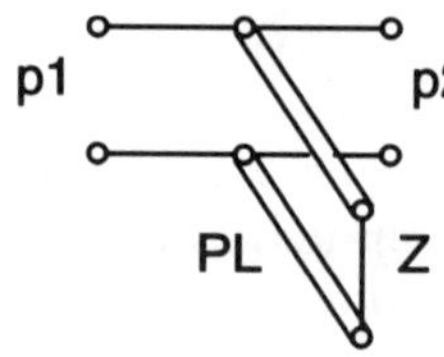

Data

p1 = Port number
p2 = Port number
Z = Characteristic impedance
PL = Physical length
K = Effective dielectric constant
A = Attenuation constant
F = Frequency for attenuation scaling

Syntax

SCPSTP p1 p2 Z = x1 PL = x2 K = x3 A = x4 F = x5

Example

SCPSTP 11 12 Z = 50 PL = 12 mm K = 2.3 A = .3 dB/m
F = 1 GHz

Note

Attenuation A is constant with frequency if F is not defined in the SCPSTP statement or if F is zero. Otherwise, A is computed as frequency dependent as

$A(f) = A$ for F = 0 or undefined

$A(f) = A * \sqrt{f/F}$ for $F \neq 0$

D.1.4.14 SCSST—Ideal Short-Circuited Series Stub

Schematic

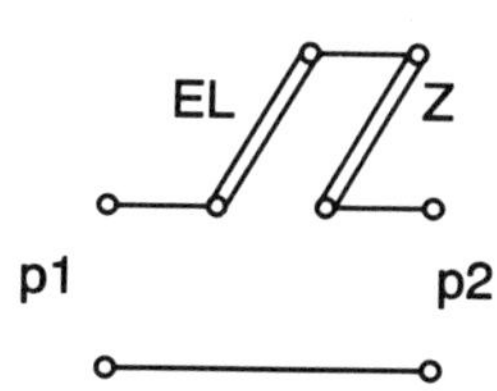

Data

p1 = Port number

p2 = Port number
Z = Characteristic impedance
EL = Electrical length
F = Frequency at which electrical length is EL

Syntax

SCSST p1 p2 Z = x1 EL = x2 F = x3

Example

SCSST 11 12 Z = 50 EL = 90 deg F = 1 GHz

Note

SCSST is an ideal short-circuited series stub.

D.1.4.15 SCSSTP—Short-Circuited Series Stub Physical Model

Schematic

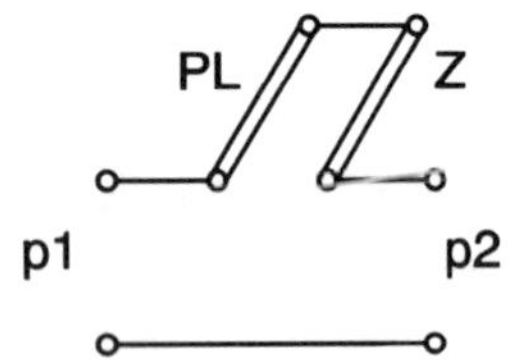

Data

p1 = Port number
p2 = Port number
Z = Characteristic impedance
PL = Physical length
K = Effective dielectric constant
A = Attenuation constant
F = Frequency for attenuation scaling

Syntax

SCSSTP p1 p2 Z = x1 PL = x2 K = x3 A = x4 F = x5

Example

SCSSTP 11 12 Z = 50 PL = 12 mm K = 2.3 A = .3 dB/m F = 1 GHz

Note

Attenuation A is constant with frequency if F is not defined in the SCSSTP statement or if F is zero. Otherwise, A is computed as frequency dependent as
$A(f) = A$ for $F = 0$ or undefined
$A(f) = A * \sqrt{f/F}$ for $F \neq 0$

D.1.4.16 TLIN—Ideal Transmission Line

Schematic

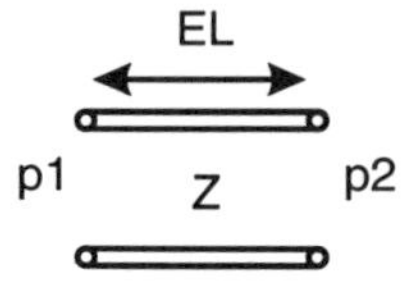

Data

p1 = Port number
p2 = Port number
Z = Characteristic impedance
EL = Electrical length in radians or in angle units
F = Frequency at which electrical length is EL

Syntax

TLIN p1 p2 Z = x1 EL = x2 F = x3

Example

TLIN 11 12 Z = 50 EL = 90 deg F = 3 GHz

Note

TLIN is an ideal lossless transmission line.

D.1.4.17 TLINP—Transmission Line Physical Model

Schematic

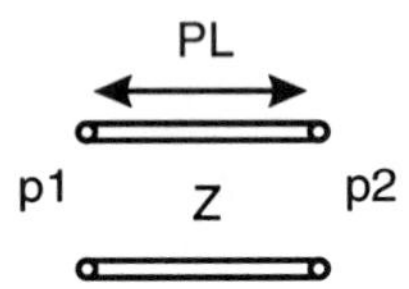

Data

p1 = Port number
p2 = Port number
Z = Characteristic impedance
PL = Physical length
K = Effective dielectric constant
A = Attenuation constant
F = Frequency for attenuation scaling

Syntax

TLINP p1 p2 Z = x1 PL = x2 K = x3 A = x4 F = x5

Example

TLINP 11 12 Z = 50 PL = 12 mm K = 2.3 A = .3 dB/m F = 1 GHz

Note

Attenuation A is constant with frequency if F is not defined in the TLINP statement or if F is zero. Otherwise, A is computed as frequency dependent as

A(f) = A for F = 0 or undefined

A(f) = A $*\sqrt{f/F}$lx for F ≠ 0

D.1.5 Microstrip Distributed Elements

MCLIN Microstrip coupled lines (EL66).
MLIN Microstrip line (EL64).
MLOC Microstrip line open circuit (EL65).
MLSC Microstrip line short circuit (EL71).

D.1.5.1 MCLIN—Microstrip Coupled Lines

Schematic

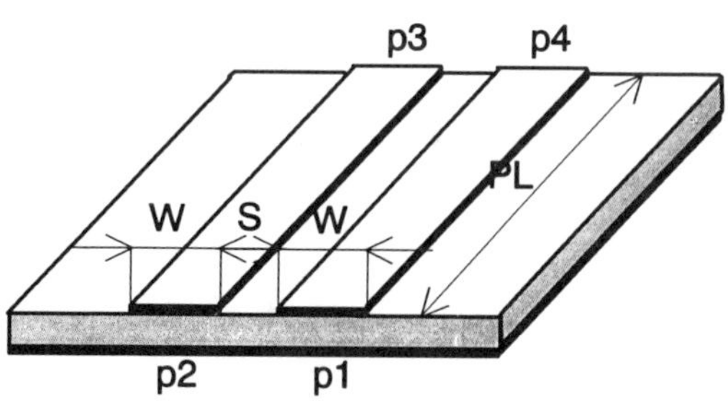

Data

p1 = Port number
p4 = Port number
W = Strip width
S = Gap width
L = Strip length

Syntax

MCLIN p1 p4 W = x1 S = x3 L = x3

Example

MCLIN 1 4 W = 1.2 mm S = 0.1 mm L = 125 mm

Note

Parameters of the MCLIN substrate is supplied by an MSUB statement, so it must be located before the MCLIN statement in the CKT block of the input data.

D.1.5.2 MLIN—Microstrip Line

Schematic

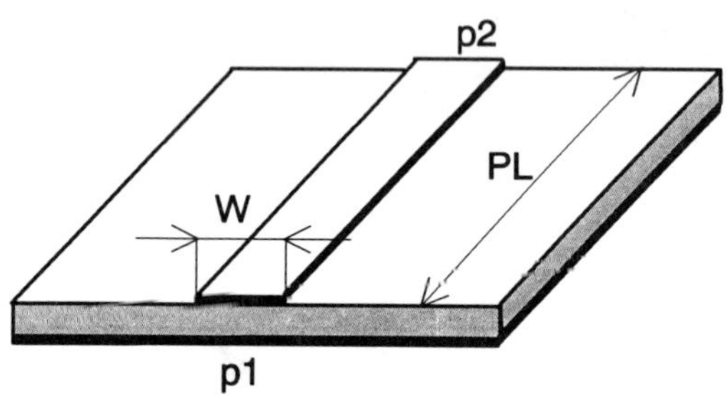

Data

p1 = Port number
p2 = Port number
W = Strip width
L = Strip length

Syntax

MLIN p1 p2 W = x1 L = x2

Example

MLIN 1 2 W = 1.2 mm L = 125 mm

Note

Parameters of the MLIN substrate are supplied by an MSUB statement, so it must be located before the MLIN statement in the CKT block of the input data.

D.1.5.3 MLOC—Microstrip Line Open Circuit

Schematic

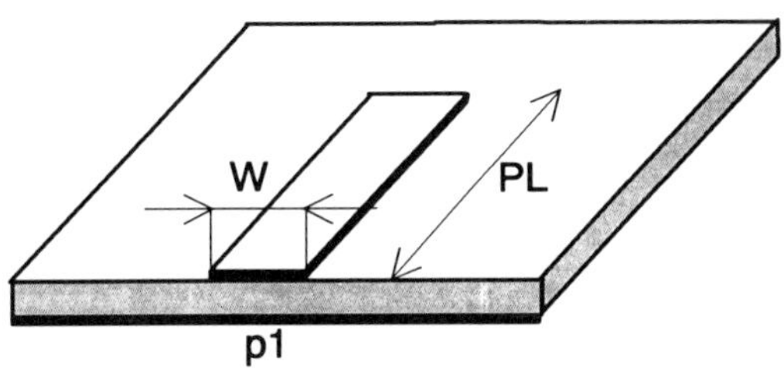

Data

p1 = Port number
W = Strip width
L = Strip length

Syntax

MLOC p1 W = x1 L = x2

Example

MLOC 1 W = 1.2 mm L = 125 mm

Notes

1. Microstrip line with an ideal open circuit at one end.
2. Parameters of the MLINE substrate are supplied by an MSUB statement, so it must be located before the MLIN statement in the CKT block of the input data.

D.1.5.4 MLSC—Microstrip Line Short Circuit

Schematic

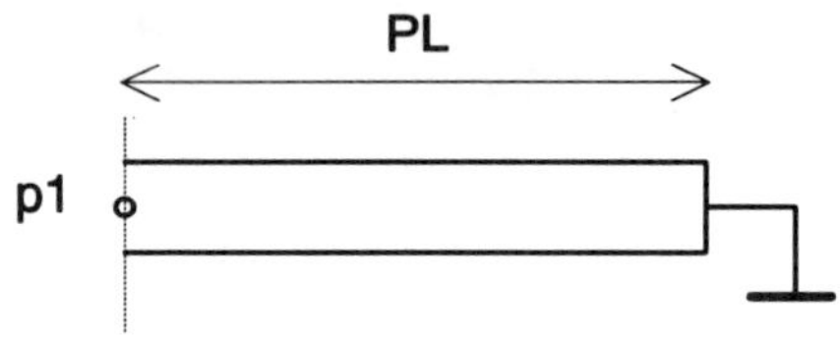

Data

p1 = Port number
W = Microstrip line width
L = Microstrip line length

Syntax

MLSC p1 W = x1 L = x2

Example

MLSC 12 W = 2.5 mm L = 125 mm

Notes

1. It is assumed that one end of a microstrip line is connected to the ground through an ideal short circuit.
2. Range of usage: $0.01 \leq W/H \leq 100$ and $\varepsilon_r \leq 128$.
3. MSUB statement specifying substrate parameters must be placed in the CKT block before the MSLC statement.

D.1.6 Microstrip Lumped Elements

RIBBON	Flat wire (EL76).
TFC	Thin film capacitor in microstrip (EL73).
TFR	Thin film resistor in microstrip (EL72).
WIRE	Round wire (EL75).

D.1.6.1 RIBBON—Flat Wire

Schematic

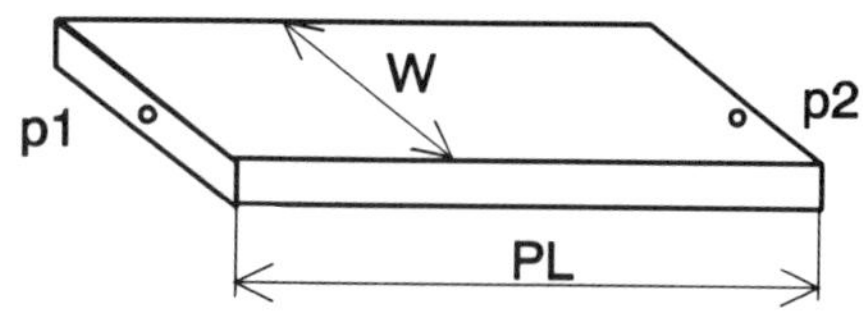

Data

p1	= Port number
p2	= Port number
W	= Wire width
PL	= Wire length
T	= Wire thickness
RHO	= Conductor resistivity normalized to that of gold

Syntax

```
RIBBON p1 p2 W=x1 PL=x2 T = x3 RHO=x4
```

Example

```
RIBBON 8 9 W= 1mm PL=10mm T = 0.3 MM RHO = .9
```

Notes

1. The JADMIC2 model for a round wire is a series lumped inductance-resistance calculated from physical dimensions.
2. The value of gold resistivity is equal to 2.44 μΩ·cm. The actual value of metal resistivity used in RIBBON computations is this value 2.44 μΩ·cm multiplied by the value assigned to RHO.

D.1.6.2 TFC—Thin Film Capacitor

Schematic

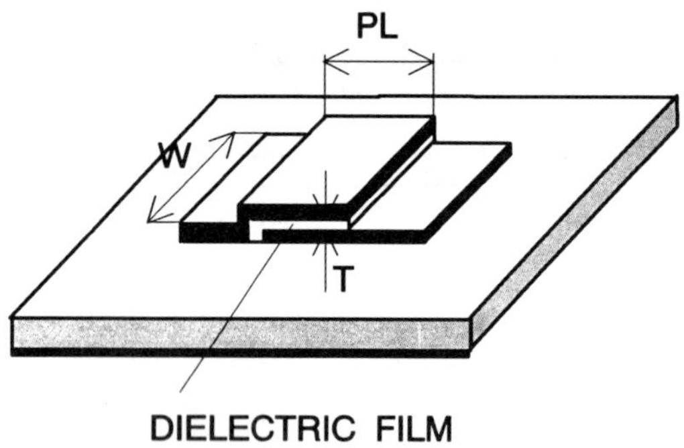

Data

p1 = Port number
p2 = Port number
W = Width
PL = Length
T = Thickness of dielectric
ER = Relative dielectric constant
RHO = Bulk resistivity of film conductor normalized to that of gold
TAND = Dielectric loss tangent

Syntax

TFC p1 p2 W = x1 PL = x2 T = x3 ER = x4 RHO = x5 TAND = x6

Example

TFC 1 2 W = 2.5mm PL = 3.mm T = .1 mm ER = 2.3 RHO = 1.
TAND = .0001

D.1.6.3 TFR—Thin Film Resistor in Microstrip

Schematic

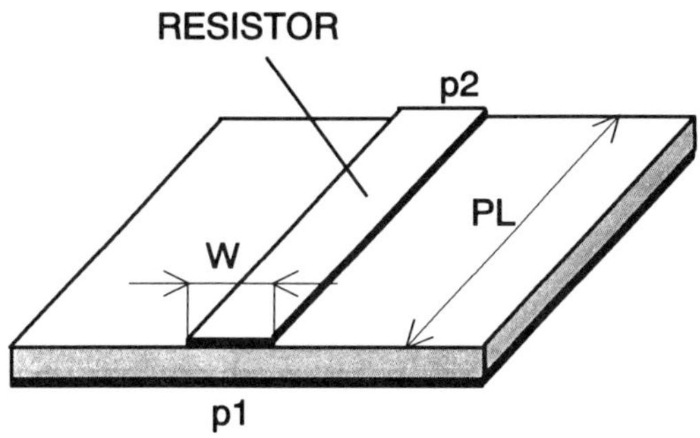

Data

p1 = Port number
p2 = Port number
W = Strip width
PL = Metal strip length
RS = Sheet resistivity Ω/□
F = Reference frequency for RS

Syntax

TFR p1 p2 W = x1 PL = x2 RS = x3 F = x4

Example

TFR 4 5 W = .8 mm PL = 17 mm RS = 100 F = 1 GHz

Notes

1. If F = 0 or if F is not defined in the TRF statement, then RS is assumed to be constant with frequency. If F ≠ 0, then RS is scaled as the square root of the frequency according to the formula

   $RS(f) = RS(F) * \sqrt{f/F}$

   in which f is the current frequency. In most practical cases, F should be assigned to 0.
2. An MSUB statement that defines substrate parameters should appear in the BLK block before the TFR statement. Only in such a case, TFR is treated as

a distributed resistor whose resistance per unit length is

R/L = RS(f)/W

3. Without MSUB preceding the TFR statement, TFR is a lumped resistor of value

r = RS(f) * L/W

D.1.6.4 WIRE—Round Wire

Schematic

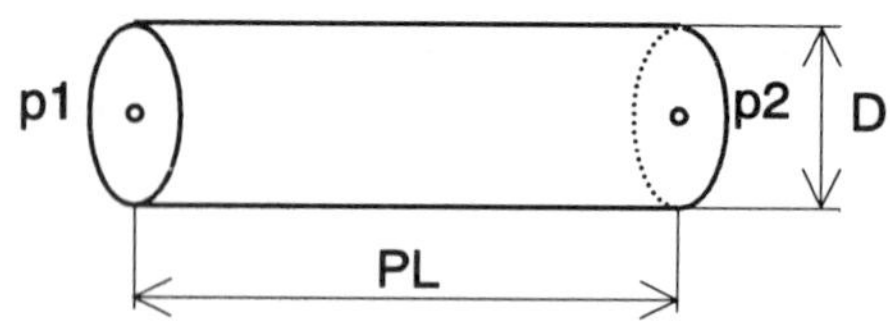

Data

p1 = Port number
p2 = Port number
D = Wire diameter
PL = Wire length
RHO = Conductor resistivity normalized to that of gold

Syntax

WIRE p1 p2 D=x1 PL=x2 RHO=x3

Example

WIRE 8 9 D= 1mm PL=10mm RHO = .9

Notes

1. The JADMIC2 model for a round wire is a series lumped inductance-resistance calculated from physical dimensions.
2. The value of gold resistivity is equal to 2.44 μΩ·cm. The actual value of metal resistivity used in WIRE computations is this value 2.44 μΩ·cm multiplied by the value assigned to RHO.

D.1.7 Universal Elements

CIR3	Ideal three-port circulator (EL86).
DELAY	Time delay element (EL87).
GAIN	Gain element (EL88).
GYR	Gyrator (EL89).
ISOLAT	Ideal isolator (EL90).
MATCH	Matched load (EL91).
OPEN	Open (EL92).
PHASE	Phase shift element (EL93).
PTEE	Parallel Tee junction (EL24).
SHORT	Short (EL94).
STEE	Series Tee junction (EL26).

D.1.7.1 CIR3—Ideal Three-Port Circulator

Schematic

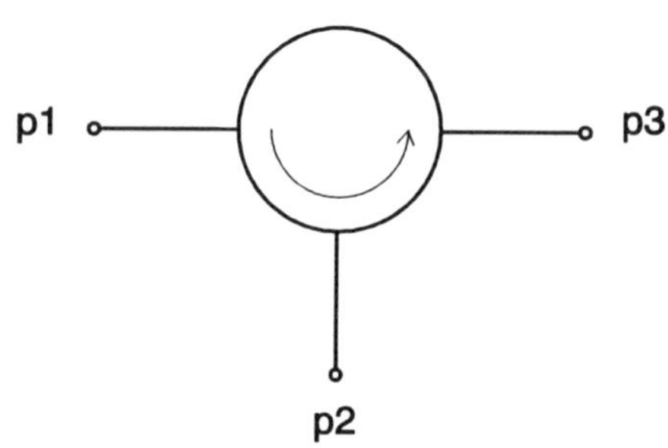

***S** matrix*

$$\mathbf{S} = \begin{bmatrix} 0 & 0 & 1 \\ 1 & 0 & 0 \\ 0 & 1 & 0 \end{bmatrix}$$

Data

p1 = Port number
p3 = Port number

Syntax

CIR3 p1 p3

Example

CIR3 1 3

D.1.7.2 DELAY—Time Delay Element

Schematic

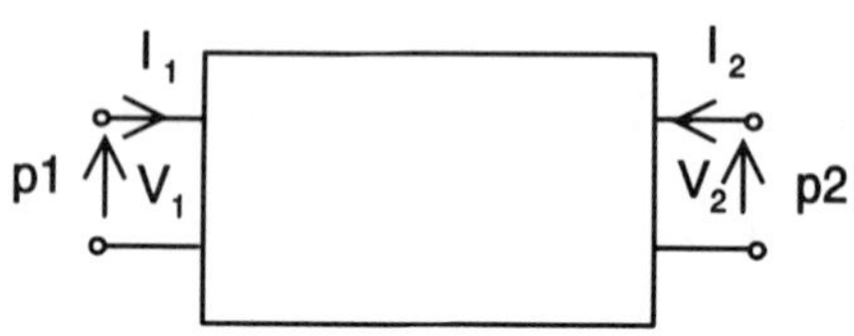

Data

p1 = Port number
p2 = Port number
T = Time delay

Syntax

DELAY p1 p2 T=x1

Example

DELAY 1 2 T = 10 ms

Notes

1. *S* parameters of the DELAY element are
   S11 = 0   S22 = 0,
   S12 = exp(−j*2*Π*f*T),   S21 = exp(−j*2*Π*f*T)
2. Any number of DELAY elements can be cascaded to obtain a desired shape as a function of frequency.

D.1.7.3 GAIN—Gain Element

Schematic

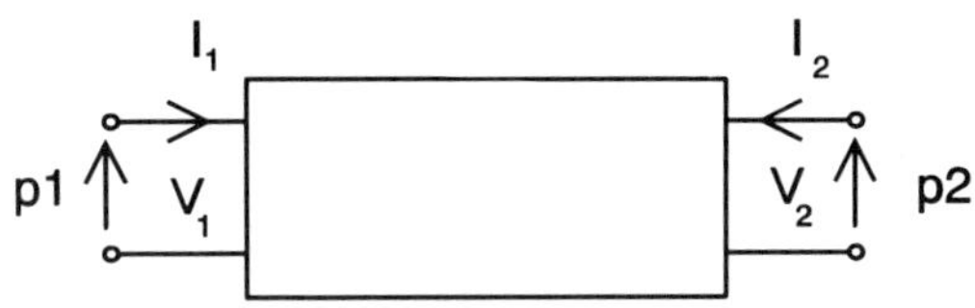

Voltage current relations

$V_2/V_1 = A(f)$	$A(f)\ [dB] = A,$	$f < F$
$I_2/I_1 = A(f)$	$A(f)\ [dB] = A + S * \log_2(f/F)$	$f > F$

Scattering matrix

$$\mathbf{S} = \begin{bmatrix} \mathbf{0} & \mathbf{1/A(f)} \\ \mathbf{A(f)} & \mathbf{0} \end{bmatrix}$$

where

f = Current frequency
A = Constant gain in decibels
S = Gain slope in dB/octave
F = Frequency at which slope starts

Data

p1 = Port number
p2 = Port number
A = Constant gain in decibels
S = Gain slope in dB/octave
F = Frequency at which slope starts, F > 0

Syntax

```
GAIN p1 p2 A = x1 S = x2 F = x3
```

Example

GAIN 3 4 A = 10 S = 6 F = 2.5 GHz

D.1.7.4 GYR—Gyrator

Schematic

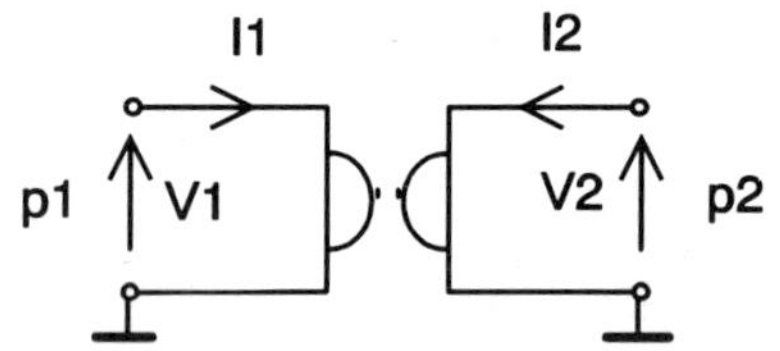

Voltage-current relationship

$V_1 = R\,I_2$
$V_2 = -R\,I_1$

Scattering matrix

$$\mathbf{S} = \begin{bmatrix} \dfrac{r^2-1}{r^2+1} & \dfrac{2r}{r^2+1} \\ \dfrac{-2r}{r^2+1} & \dfrac{r^2-1}{r^2+1} \end{bmatrix} \quad \text{where } r = R/Z_R$$

Data

p1 = Port number
p2 = Port number
R = Gyrator ratio

Syntax

GYR p1 p2 R = x1

Example

GYR 6 7 R = 25

D.1.7.5 ISOLAT—Isolator

Schematic

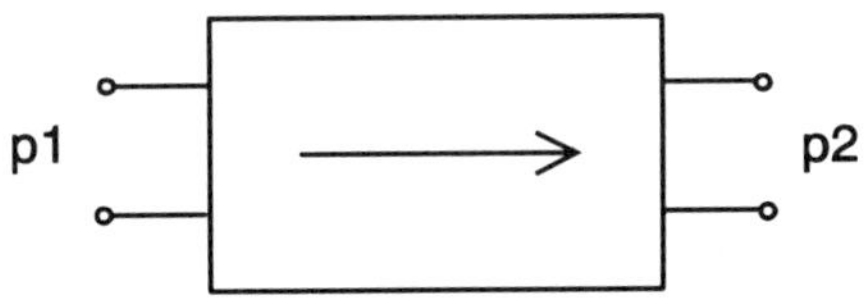

***S** matrix*

$$\mathbf{S} = \begin{bmatrix} 0 & 0 \\ 1 & 0 \end{bmatrix}$$

Data

p1 = Port number
p2 = Port number

Syntax

ISOLAT p1 p2

Example

ISOLAT 23 24

D.1.7.6 MATCH—Matched Load

Schematic

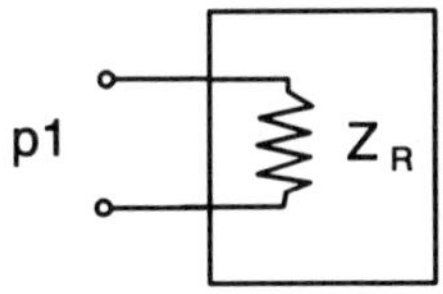

Data

p1 = Port number

Syntax

MATCH p1

Example

Match 45

D.1.7.7 OPEN—Open Circuit

Schematic

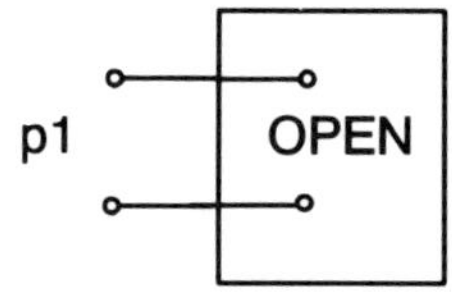

Data

p1 = Port number

Syntax

OPEN p1

Example

OPEN 21

D.1.7.8 PHASE—Phase Shift Element

Schematic

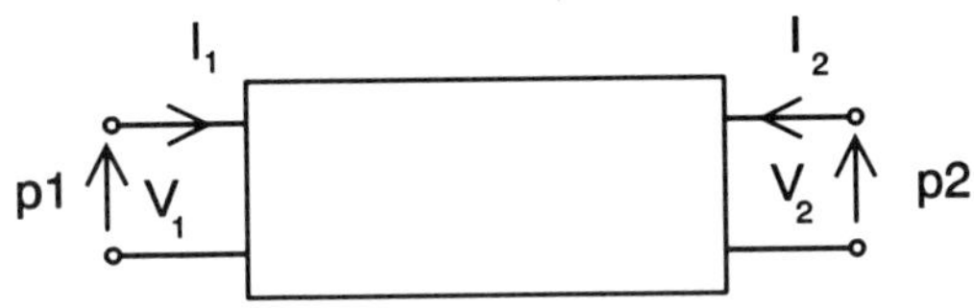

Voltage-current relationships:

$$V_2/V_1 = \exp(j\ ANG(f))$$
$$I_2/I_1 = \exp(j\ ANG(f))$$

***S** matrix*

$$\mathbf{S} = \begin{bmatrix} 0 & \exp(-j\ ANG(f)) \\ \exp(j\ ANG(f)) & 0 \end{bmatrix}$$

where

$ANG(f) = A \qquad f < F$
$ANG(f) = A + S \log_2 (f/F) \qquad f > F$
f = Current frequency
A = Constant phase
S = Phase slope per octave
F = Frequency at which slope starts.

Data

p1 = Port number
p2 = Port number
A = Constant phase, $-180° \leq A \leq 180°$
S = Slope per octave
F = Frequency at which slope starts, $F > 0$

Syntax

PHASE p1 p2 A = x1 S = x2 F = x3

Example

PHASE 7 8 A = 45 S = 20 F = 1.4 GHz

D.1.7.9 PTEE—Parallel Tee Junction

Schematic

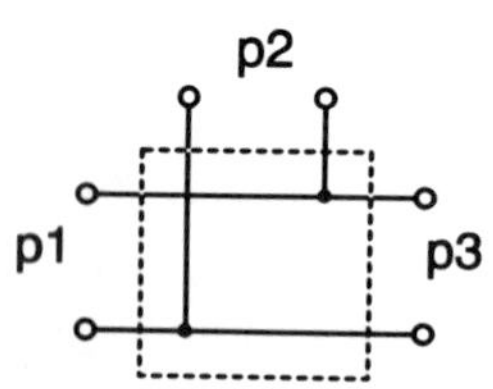

Data

p1 = Port number
p3 = Port number

Syntax

PTEE p1 p3

Example

PTEE 1 3

D.1.7.10 SHORT—Short Circuit

Schematic

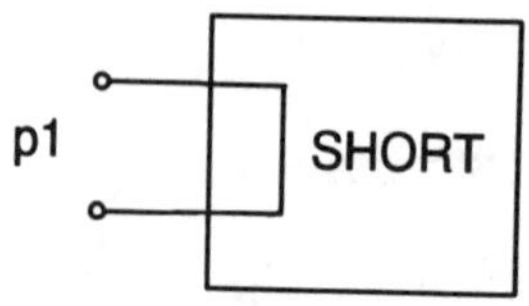

Data

p1 = Port number

Syntax

SHORT p1

Example

SHORT 7

D.1.7.11 STEE—Series Tee Junction

Schematic

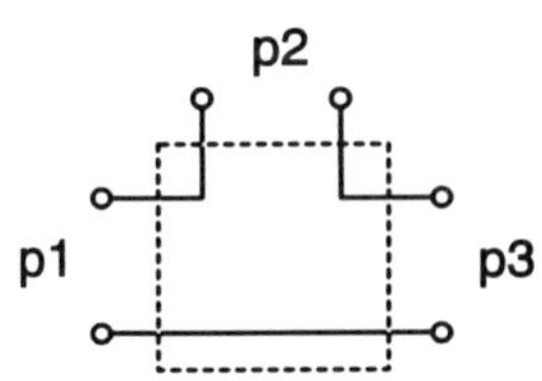

Data

p1 = Port number
p3 = Port number

Syntax

STEE p1 p3

Example

STEE 1 3

D.2 DEFINING SUBSTRATE PARAMETERS OF MIC ELEMENTS

In the case of microwave integrated circuits simulation, the MSUB statements are used in the CKT block to define substrate parameters for JADMIC2 MIC elements. The TAND defines the loss tangent of a dielectric substrate.

MSUB Statement defining substrate parameters of microstrip elements.
TAND Statement defining dielectric loss tangent for microstrip substrate.

D.2.1 MSUB—Microstrip Substrate Parameters

The MSUB statement defines electrical and geometrical parameters of the substrate for one or more microstrip elements.

Physical layout

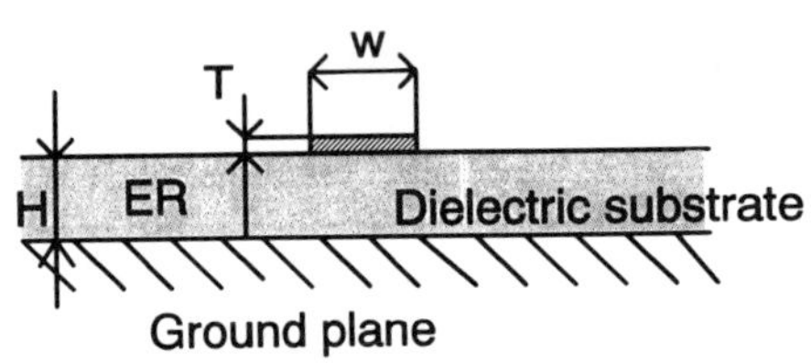

Data

ER = Substrate relative dielectric constant
H = Substrate thickness
T = Metal thickness
RHO = Metal bulk resistivity normalized to that of gold
RGH = rms surface roughness

Syntax

MSUB ER = x1 H = x2 T = x3 RHO = x4 [RGH = x5]

Example

MSUB ER = 2.3 H = 1. MM T = 0.05 mm rho = 1

Notes

1. MSUB must be included in the CKT block before any statement that defines the microstrip element.
2. It is assumed that the conductor metal of the strip is a single metal layer with bulk resistivity RHO.
3. JADMIC uses 2.44 E-6 Ω/cm for the bulk resistivity of gold. The actual value of the resistivity of microstrip metal is this value multiplied by RHO. Conductor losses are neglected if RHO = 0.
4. Not all parameters of a microstrip substrate must be specified in the MSUB statement. Specifying RGH overrides default values of this parameter. The default value for RGH is 0.
5. The width of the conductor strip and the length of the microstrip elements are defined in statements that define these elements.

In the case of a microstrip structure using electrodeposited or rolled copper laminates, the thickness of the metal layer is specified by a manufacturer in "ounces of copper." Manufacturers of rolled copper laminates give the weight of copper deposited on one square foot of laminate surface. The relationship between the weight and the thickness of copper is

Weight	Thickness	
(oz)	(in)	(mm)
0.5	0.0007	0.01778
1.0	0.00135	0.03429
2.0	0.0028	0.07112

D.2.2 TAND—Dielectric Loss Tangent

The TAND statement is used to specify the dielectric loss tangent for dielectric substrates of any MMIC or MIC distributed elements.

Data

TAND = Dielectric loss tangent

Syntax

TAND = x1

Example

TAND = 0.0001

Notes

1. JADMIC2 assumes loseless dielectric substrates for all MIC and MMIC network elements ($\tan \delta = 0$).
2. The TAND statement assigns a value to $\tan \delta$ and overrides the default value $\tan \delta = 0$.
3. The CKT block may contain any number of TAND statements. The value assigned to $\tan \delta$ remains active until replaced by a subsequent TAND statement.

APPENDIX E

Describing Topology of the Network—TPL Block

Once the circuit elements have been defined and assigned their port numbers and numerical values of element parameters, the next step is to interconnect them in the desired manner, which is accomplished by means of special circuit topology description key words located in the TPL block of the input file. The format and the meaning of the different key words are given later in this section.

The generator port number and the load port number must be defined in the TPL section, and this is accomplished by using two commands, starting with mandatory words GEN and LOAD, respectively. The generator and the load may be any one one-port element described in Appendix C, namely, SRLC, PRLC, Z1PORT, Y1PORT, S1PORT, MATCH, or RES. Hence, generator and load reflection coefficients may be defined as frequency dependent.

The JADMIC2 key words that define the topology of the analyzed circuit are arranged into groups.

1. Key words that define connections between ports of circuit elements are:

CON Connection of two ports.
STEE Series connection of three ports.
PTEE Parallel connection of three ports.
CROSS Parallel connection of four ports.

2. Key words that define the signal generator port and the load port are:

GEN Generator port.
LOAD Load port.

Using these statements the user defines the input and output port of the circuit to be analyzed, all circuit functions are referenced to these ports.

E.1 DEFINING CONNECTIONS BETWEEN PORTS OF NETWORK ELEMENTS

E.1.1 CON—Connection of Two Ports

The CON command is used to define junctions of network ports that are connected. Numbers of adjacent ports of network elements must be grouped in pairs by the equality sign.

Syntax

CON p1 = p2 p3 = p4 . . .

Example

CON 1=2 3=4 5=6

E.1.2 PTEE—Parallel Tee Connection

Although most typical microwave circuits can be defined by the standard two-port type connections, there are cases when additional flexibility in defining circuit topology is desired. The PTEE command is used to define circuit topology when three ports are connected in parallel.

Syntax

PTEE p1=p2=p3 p4=p5=p6 . . .

Example

PTEE 1=2=3 13=23=34

E.1.3 STEE—Series Tee Connection

Although most typical microwave circuits can be defined by the standard two-port type connections, there are cases when additional flexibility in defining circuit topology is desired. The STEE command is used to define circuit topology when three ports are connected in series, respectively.

Syntax

STEE p1=p2=p3 p4=p5=p6 . . .

Example

STEE 2=5=45 8=21=33

E.1.4 CROSS—Parallel Connection of Four Ports

The CROSS statement is used to define circuit topology when four ports are connected in parallel.

Syntax

CROSS p1=p2=p3=p4 p5=p6=p7=p8 . . .

Example

CROSS 2=6=12=16 5=7=21=22

E.2 DEFINING THE SIGNAL GENERATOR PORT AND THE LOAD PORT OF THE NETWORK

The GEN statement of the TPL block is used to assign the role of the signal generator to the particular one-port element of the analyzed circuit. The signal generator exciting the analyzed circuit may be any one-port type elements such as SRLC, PRLC, Z1PORT, Y1PORT, S1PORT, MATCH, or RES.

The LOAD statement of the TPL block is used to assign the role of the load to the particular one-port element of the analyzed circuit. The load of the circuit may be any of the one-port type circuit elements such as SRLC, PRLC, Z1PORT, Y1PORT, S1PORT, MATCH, or RES.

E.2.1 GEN—Signal Generator Port

The GEN command located in the TPL block identifies the signal generator port.

Syntax

GEN = p1

Example

GEN = 5

E.2.2 LOAD—Load Port

The LOAD statement identifies the load port in the circuit.

Syntax

LOAD = p1

Example

LOAD 36

The following is an example of the TPL block:

```
TPL
  CON 1=2 3=4 5=6 7=8
  GEN 1
  LOAD 8
END
```

APPENDIX F

Specifying a Set of Frequencies—FREQ Block

The frequency block starts with the mandatory key word FREQ. All frequency points relating to analysis and to optimization must be defined in this block. Subsets of frequencies that are used in optimization of the network are then specified in their respective blocks.

Three mnemonics—the key words FSET, FLIN, and FLOG—are used to define discrete frequency points in the JADMIC2 program.

F.1 FSET—A SET OF FREQUENCY POINTS

The FSET mnemonic is used to define any number of discrete frequencies f_1, $f_2, \ldots, f_n$. The format of this command is

$$\text{FSET } \mathrm{f}_1, \mathrm{f}_2, \ldots, \mathrm{f}_n$$

Example

FSET 1.0 MHZ 10. MHZ 1 GHZ 10. GHZ

F.2 FLIN—LINEARLY DISPOSED FREQUENCY POINTS

The second mnemonic FLIN is used to generate a list of n linearly stepped frequencies in the range f_1, f_n. The format of this command is

$$\text{FLIN } f_1 \; f_n \; n$$

FLIN generates the following set of frequencies:

$$f = \{f_1, f_1 + \Delta f, f_1 + 2\Delta f, \ldots, f_1 + (n-1)\,\Delta f\}$$

$$\text{where } \Delta f = (f_n - f_1)/(n-1).$$

Example

FLIN 1 GHZ 10 GHZ 10

F.3 FLOG—LOGARITHMICALLY DISPOSED FREQUENCY POINTS

The mnemonic FLOG generates a list of logarithmically stepped frequencies. The FLOG statement has the format

$$\text{FLOG } f_1 \quad f_n \quad n$$

and the set of frequencies is

$$f = \{f_1, f_1 \cdot r, f_1 \cdot r^2, f_1 \cdot r^3, f_n = f_1 \cdot r^{(n-1)}\}$$

where $r = 10^{(n-1)^{-1} \log_{10}(f_n/f_1)}$.

Example

```
FREQ
  FLIN 10 MHZ 50 MHZ 5 FLOG 100 MHZ 1000 MHZ 10
  FSET 2.0 GHZ 5 GHZ
END
```

This FREQ data block specifies five linearly stepped frequencies in the range 10 MHz to 50 MHz, ten exponentially stepped frequencies in the range 100 MHz to 1000 MHz, and two discrete frequencies 2 GHz and 5 GHz.

APPENDIX G

Creating and Using Library Blocks—LIB Block

Measured multiports such as S1PORT and S2PORT require measured scattering parameter data. A library block, starting from the mandatory key word LIB, is used to define such parameters. The data contained in the LIB block of an input file are referenced individually by the label given by the user to the LIB block. Each of the LIB blocks of the input file starts with its own name, the label, and can contain several lines of data. Its internal organization depends upon the type of data it contains and the type of statements that reference the data.

G.1 STRUCTURE OF THE LIB BLOCK SPECIFYING MEASURED SCATTERING PARAMETERS

The measured scattering parameter data that is required by measured multiports such as S1PORT and S2PORT must be placed in a LIB block as follows.

1. The first entry after the LIB mnemonic must be a data name (a label—maximum six alphanumeric characters long). The names used as labels for a LIB block might not be the JADMIC2 mnemonics used as key words to define, for example, element types, topology, and output data.
2. The second entry defines the type of element to be described by the measured or calculated data (e.g., S1PORT or S2PORT).
3. The reference impedance ($ZR = r$) for the scattering matrix parameters.
4. The next entries are numbers defining scattering parameters for each frequency.

G.1.1 Specifying S1PORT Type Element Scattering Parameters

For the S1PORT type elements the order is frequency and S11 (two numbers—the first one representing magnitude, and the second one defining the phase in degrees).

The syntax of the LIB block for the S1PORT element is

```
LIB
label  S1PORT     [ZR = x1]
f1     mod {S11}  arg {S11}
⋮         ⋮          ⋮
fn     mod {S11}  arg {S11}

END
```

For easy reading of the input file it is recommended that the first three entries be set on one line and the next data—f_k, mod{S_{11}}, and arg{S_{11}}—into separate lines of the LIB block.

Example

```
LIB
Q12  S1PORT  ZR = 50 OHM
  1.0 GHZ     0.12     35.
  1.5 GHZ     0.15     23.
  2.0 GHZ     0.20     15.
  2.5 GHZ     0.27      6.
END
```

G.1.2 Specifying S2PORT Type Element Scattering Parameters

For S2PORT type elements the format for presenting data to JADMIC in the LIB block is

```
LIB
label  S2PORT  [ZR = r]

  f1 mod {S11} arg {S11} mod {S12} arg {S12} mod {S21} arg {S21} mod {S22} arg {S22}
```

. . . .
. . . .
. . . .

f_n mod $\{S_{11}\}$ arg $\{S_{11}\}$ mod $\{S_{12}\}$ arg $\{S_{12}\}$ mod $\{S_{21}\}$ arg $\{S_{21}\}$ mod $\{S_{22}\}$ arg $\{S_{22}\}$
END

Example

```
LIB
FET1   S2PORT   ZR = 60.   OHM
  10.  GHZ  0.42  35.  0.01  52.  2.1  145.  0.21  -140.
  10.2 GHZ  0.45  38.  0.011 55.  2.   150.  0.20  -134.
  10.4 GHZ  0.51  47.  0.015 61.  1.97 157.  0.19  -125.
  10.6 GHZ  0.50  49.  0.017 67.  1.91 163.  0.17  -120.
END
```

G.2 SPECIFYING NOISE PARAMETERS OF S2PORT TYPE ELEMENT

In the case of two-port elements (transistors) defined by means of their frequency-dependent scattering parameters, a LIB block may be used to define frequency-dependent noise parameters. Noise parameters of a two-port are minimum noise figure F_{min}, optimum (with respect to noise figure) reflection coefficient Γ_{opt} of the signal generator port, and the equivalent noise resistance R_n normalized with respect to the reference impedance ZR, $r_n = R_n/ZR$.

The data describing noise properties of S2PORT type element must be placed inside a given group of data referenced by its level and located just immediately after the scattering parameter data.

The first entry following scattering parameters data must be the key word NOISE. The next entries must be numbers that define the noise parameters of the two-port at each frequency. The format for the LIB group that defines noise parameters is

LIB
libname S2port [ZR = r]
Scattering parameters
NOISE
f_1 F_{min} mod$\{\Gamma_{g\ opt}\}$ arg$\{\Gamma_{g\ opt}\}$ r_n
f_2 F_{min} mod$\{\Gamma_{g\ opt}\}$ arg$\{\Gamma_{g\ opt}\}$ r_n
f_3 F_{min} mod$\{\Gamma_{g\ opt}\}$ arg$\{\Gamma_{g\ opt}\}$ r_n
. . . .

```
. . . . . . . . . . . . . . . . . . . . . . . . . .
. . . . . . . . . . . . . . . . . . . . . . . . . .
```

f_n F_{min} mod{ $\Gamma_{g\ opt}$ } arg{ $\Gamma_{g\ opt}$ } r_n
END

The following is the example of a LIB block that specifies scattering parameters and noise parameters of the S2PORT element called FET2.

```
LIB
  FET2 S2PORT ZR = 50 OHM
  10.  GHZ  0.42  35.  0.01  52.  2.1  145.  0.21  -140.
  10.2 GHZ  0.45  38.  0.011 55.  2.   150.  0.20  -134.
  10.4 GHZ  0.51  47.  0.015 61.  1.97 157.  0.19  -125.
  10.6 GHZ  0.50  49.  0.017 67.  1.91 163.  0.17  -120.
NOISE
  10.  GHZ  1.42  0.45  52.   .38
  10.2 GHZ  1.45  0.43  60.7  .20
  10.4 GHZ  1.51  0.40  74.0  .20
  10.6 GHZ  1.60  0.39  78.5  .40
END
```

APPENDIX H

Optimizing Circuits and Defining Optimization Goals—OPT Block

In addition to network analysis, the user can request the optimization of specified parameters to approximate a desired performance of the network. A maximum of 30 parameters can be varied simultaneously. The error functions are user defined in the OPT block of the input file by means of proper key words and coefficients.

As in the case of all other input blocks, the optimization block starts with its own key word, here OPT, and ends with the END statement.

H.1 OPTIMIZATION METHOD USED IN JADMIC2

JADMIC2 uses the Powell optimization method that iteratively arrives at the new parameter values at which specifications of the design are met.

The method may use two different objective functions. They are the least-squares objective function and the generalized least *p*th objective function.

H.1.1 Powell Method

The Powell method implemented in JADMIC2 is of direction-set type optimization methods. It does not require computation of the objective function gradient to find a direction for the search of the objective function minimum.

H.2 FORMULATION OF THE OBJECTIVE FUNCTIONS

The objective functions used in JADMIC2 for microwave network optimization are formulated as l_2 norm or l_p norm of the magnitudes of all error functions for each specification of the designed microwave network.

H.2.1 Least-Squares Objective Function

The least-squares objective function is a scalar function U equal to the l_2 norm of the magnitudes of all error functions e, which are the differences between the actually calculated network responses F and the given specifications (goals) S.

To understand the objective function computation, let S_{ij} represents the ith specification (goal) considered within the jth frequency range. The contribution of the network response function R_{ij} to the scalar objective function U can be computed as

$$U_{ij} = W_{ij} * \sum_k e_{ij}(f_k)$$

where W_{ij} is the weighting factor, $e_{ij}(f_k)$ is the error function calculated at a frequency f_k belonging to the jth frequency range F_j. The summation is performed over all frequencies in F_j.

Computing the error functions depends on the type of the response specifications. For the "greater than," GT, specification

$$e_{ij}(f) = \begin{cases} |R_{ij}(f) - S_{ij}|^2 & \text{if } R_{ij}(f) < S_{ij} \\ 0 & \text{if } R_{ij}(f) > S_{ij} \end{cases}$$

For the "equal to," EQ, specification

$$e_{ij}(f) = | R_{ij}(f) - S_{ij} |^2$$

For the "less than," LT, specification

$$e_{ij}(f) = \begin{cases} 0 & \text{if } R_{ij}(f) < S_{ij} \\ |R_{ij}(f) - S_{ij}|^2 & \text{if } R_{ij}(f) > S_{ij} \end{cases}$$

In the next step the contributions U_{ij} from all network function responses in the current frequency range are summed and the result is divided by the number of frequencies N_j that exist within the jth frequency range:

$$U_j = \frac{\sum_i U_{ij}}{N_j}$$

In the final step of computing the scalar objective function U, the contributions U_j over all frequency ranges are summed:

$$U = \sum_j U_j$$

The different weighting factors for different frequency ranges for the same network response functions can be used to introduce the greater or lesser importance of the particular ranges in relation to the other ranges of the particular response function. The different weighting factors for different network response functions within a frequency range will attach greater or lesser importance to a given response function relative to other responses within that frequency range.

To summarize, the computation of the scalar objective function can be described as the triple summation in the form

$$U = \sum_j \left\{ \left[\sum_i \left(W_{ij} \sum_k | R_{ij} - S_{ij} |^2 \right) \right] / N_j \right\}$$

in which:

- The first summation index k runs over all discrete frequencies in the frequency range j;
- The second index i indicates summation over all network response functions in the frequency range j;
- The third index j indicates summation over all frequency ranges.

It is assumed in JADMIC2 that the maximum number of frequency ranges is five.

Within each frequency range the number of response functions is limited to 20, but at the same time this number is limited to 25 within all frequency ranges.

H.2.2 Generalized Least *p*th Objective Function

In the case of the least *p*th objective function the magnitudes of individual errors at each frequency are raised to the *p*th power instead of squaring them.

The error functions $e_{ij}(f_k)$ computed with respect to the lower and upper specifications may have positive as well as negative values. This happens when a "window" specification is imposed on a network function and the function moves inside that window. The generalized least *p*th objective function allows placing the response function in the middle of two specifications. A negative value of $e_{ij}(f_k)$ means that at this frequency the specification has been fulfilled to excess.

This consideration leads to the following form of the generalized least pth objective function:

$$U = \begin{cases} \sum\limits_{k \in I^+} (e(f_k))^p & \text{for } M_e > 0 \\ 0 & \text{for } M_e = 0 \\ \dfrac{-1}{\sum\limits_{k=1}^{m} (-e(f_k))^p} & \text{for } M_e < 0 \end{cases}$$

where $M_e = \max\{e(f_k)\}$ denotes the maximum error, m the total number of frequency points in all frequency ranges, and I^+ a set of frequency points that corresponds to nonnegative error functions.

The cases $M_e > 0$ and $M_e < 0$ are clearly distinguished. For the first case the objective function is formed from the nonnegative error functions only. For the second case, where $M_e < 0$, all error functions take part in the formulation of the objective function. In this case, all errors are negative, which means that all network specifications are satisfied. This permits the minimization of the objective function even after all the specifications are met. The resulting network will oversatisfy the design specifications as much as possible.

The user may choose the value of the exponent p. With large values of p, the user obtains results that are very close to the minimax optimum.

H.3 CIRCUIT FUNCTIONS IN JADMIC2 OPTIMIZATION

The desired network response can be specified as any of the network scattering matrix parameters and as the noise figure computed with respect to the input port and to the output port of the entire network. These functions are

- Real parts of the overall network scattering parameters: RES11, RES12, RES21, RES22;
- Imaginary parts of the overall network scattering parameters: IMS11, IMS12, IMS21, IMS22;
- Magnitudes of the overall network scattering parameters: MS11, MS12, MS21, MS22;
- Phases of the overall network scattering parameters: PS11, PS12, PS21, PS22;
- Noise figure of the entire network: NF;
- Group delay: GDELAY.

In the above mnemonics, numbers 11, 12, 21, and 22 indicate parameters related to overall network input and output ports.

The JADMIC2 optimizer first compares computed network response(s) with given specification(s) and then changes network parameters to minimize the difference between the computed network function and the desired specification. The error function is used to compare computed and desired responses of the designed network.

H.4 DEFINING SPECIFICATIONS OF THE OPTIMIZED CIRCUIT FUNCTIONS

Error functions that are used in optimization arise from the difference between the actually calculated responses and the given specification.

As discussed in Chapter 7, error functions can be defined with respect to lower specifications or to upper specifications, with respect to both lower and upper specifications, and with respect to a single specification.

The definitions of error functions that are minimized generally require the following form of statement (syntax):

[FL = f_l FU = f_u] function G = g1 [{ g2 | LT | GT }] [W = w]

The key words FL and FU are used to specify a subset of the whole set of frequency points defined in the FREQ data block. FL and FU specify the lower and upper limits of a range of frequencies over which optimization of the particular network function will be performed. FL and FU may be used to define different frequency bands for different network functions. If the statement that defines the error functions does not contain a frequency specification, then all frequencies specified in the FREQ block are applied to the optimization of the particular network function.

The next key word (mnemonic) specifies the network function. Its form may be one of the following:

{ RE | IM | M | P } S { 1 | 2 } { 1 | 2 }

with no spaces between, and

NF.

Letters and digits that create the key words denote the following characteristics of S11, or of S12, or of S21, or of S22 scattering parameters of the overall network:

- RE, Real part;
- IM, Imaginary part;
- M, Magnitude;
- P, Phase.

The NF mnemonic indicates the noise figure of the overall network.

The equal sign indicates the goal whose value can be an actual ratio or expressed in decibels. The key words LT and GT indicate, respectively, lower than and greater than.

The key word W indicates the weight coefficient (optional) for the defined design specification.

The default value for W is 1.0. By specifying the weight, it is possible to emphasize one response function over others in the error function computation. Greater weight should be assigned to partial errors that have low magnitudes. For example, in the case of an amplifier design, S12 should have a greater weight than S21.

Figures H.1(a) and (b) present design specifications for two different network functions $|S_{21}|$ and $|S_{11}|$. The objective of the design is to set the magnitude of S_{21} between 9 dB and 11 dB and the magnitude of S_{11} smaller than 0.15 in the frequency band 1 GHz to 3 GHz. The statement defining these design specifications has the form

FL = 1 GHZ FU = 3 GHZ MS21 G = 9 DB 11 DB MS11 g = 0.15 LT

Another example is illustrated by Figure H.2, namely, the single specification for the $|S_{21}|$ parameter. The statements defining this specification have the form

FL = 1.0 GHZ FU = 1.5 GHZ MS21 = 0
FL = 1.5 GHZ FU = 2.5 GHZ MS21 = 10
FL = 2.5 GHZ FU = 3.0 GHZ MS21 = 0

Circuit parameters whose values are changed by the optimization procedure must have specified limits of allowable changes. As nominal values of network parameters, their allowable limits are specified by enclosing them in question marks as follows:

? lower limit nominal value upper limit ?

Lower and upper limits may be given as numerical values or, optionally, as a percentage of the initial value of the parameter under consideration.

For example, to specify the characteristic impedance Z_0 of initial values 50 Ω and 45 Ω and 60 Ω, respectively, as lower and upper limits, the specifications of this parameter in element statement should have one of the following forms:

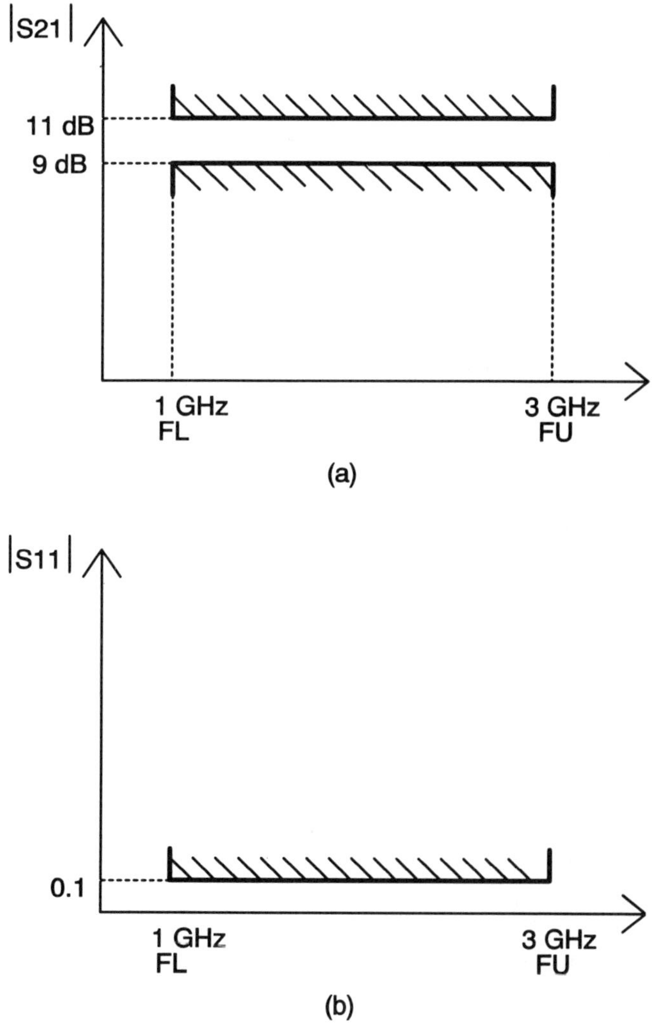

Figure H.1 (a) An example of specifications for the $|S_{21}|$ parameter. (b) An example of specifications for the $|S_{11}|$ parameter.

Z = ? 10% 50 OHM 20 % ?
Z = ? 45 50 60 ?
Z = ? 10% 50 60 ?
Z = ? 10% 50 20% ?

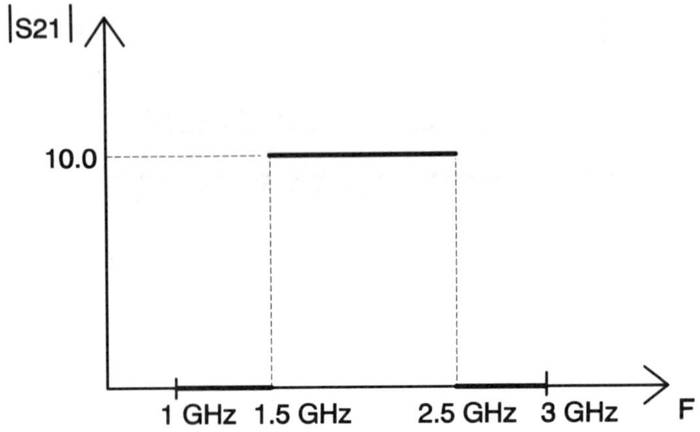

Figure H.2 An example of the single specification for the $|S_{21}|$ parameter.

H.5 SPECIFYING THE TERMINATION CRITERIA AND OPTIMIZATION PARAMETERS

The optimization block must contain a statement that defines the value of the terminating parameter FTOL. FTOL is a nonnegative variable that specifies the fractional tolerance of the objective function $U(\mathbf{p})$ such that failure to decrease by more than this amount on one iteration signals completion. Execution of network optimization is terminated.

The next parameter, called TOL, defines the fractional accuracy with which in each iteration the optimization procedure finds the optimal value of each network parameter. The TOL should not be smaller than the square root of the floating point precision of the computer.

The third optimization parameter, called ITMAX, defines the maximum number of iterations allowed.

The forth optimization parameter, PTH, is a value of p used in the least pth objective function formulation.

Summarizing, the statement that specifies optimization parameters must be of the form

TOL = x1 FTOL = x2 ITMAX = x3 [PTH = x4]

Example

TOL = 1.E-4 FTOL = 1.E-4 ITMAX = 10 PTH = 10

APPENDIX I

Specifying the Output Functions—OUT Block

JADMIC2 provides both tabular and plotted outputs. The printed output provided with JADMIC2 can be subdivided into three parts.

1. Print of the input data file (echo-check). If a syntax error is detected by the syntax analyzer, a self-explanatory error message is inserted and further reading of the input data file is terminated.
2. A detailed presentation of the microwave circuit (as it is understood by JADMIC2). Presentation of the circuit includes:

   a. Presentation of each circuit element, including the exact meaning of the numerical values given by means of key words;
   b. Description of network topology;
   c. Optimization specifications (if any).

3. The performance of the overall circuit considered as a two-port.

The results of circuit analysis and optimization may also be presented as plots in rectangular and/or circular coordinates. JADMIC2 also provides Smith charts, which display scattering parameters, input/output stability circles, constant power gain circles, and constant noise figure circles.

I.1 OUTPUT DATA OPTIONS

Output data options are selected by means of the key words used in the OUT block of the input file for JADMIC2. Desired outputs defined in the output block start with the key word OUT and ends with the key word END. All network functions and parameters are computed in the frequency band defined in the FREQ block of the input data file.

I.1.1 S—Overall Network Scattering Parameters

S is the key word that calls for the presentation of scattering parameters with respect to the input and output ports of the overall network. Scattering parameters are computed in the frequency band defined in the FREQ block.

I.1.2 Y—Overall Network Admittance Parameters

Y is the key word that calls for the computation of the admittance parameters of the overall circuit. The resultant two-port admittance matrix parameters are computed in the frequency band defined in the FREQ block.

I.1.3 Z—Overall Network Impedance Parameters

Z is the key word that calls for the computation of the impedance parameters of the overall circuit. The resultant two-port impedance matrix parameters are computed in the frequency band defined in the FREQ block.

I.1.4 NF—Overall Circuit Noise Figure

NF is the key word that calls for the computation of the noise figure of the overall circuit. The noise figure of the resultant two-port is computed in the frequency band defined in the FREQ block.

I.1.5 PGAIN—Power Gain

PGAIN is the key word that calls for the computation of the power gain. The input port is the circuit port connected to the signal generator port defined in the TPL block. Similarly, the output port of the circuit is the port connected to the load port of the circuit.

I.1.6 APGAIN—Available Power Gain

APGAIN is the key word that invokes the computation of the available power gain of the resultant two-port.

I.1.7 TPGAIN—Transducer Power Gain

TPGAIN is the key word that invokes the computation of the transducer power gain of the resultant two-port.

I.1.8 VTGAIN—Voltage Transfer Gain

VTGAIN is the key word that invokes the computation of the voltage transfer gain of the resultant two-port.

I.1.9 INLOSS—Insertion Loss

INLOSS is the key word that invokes the computation of the insertion loss of the resultant two-port.

I.1.10 GDELAY—Group Delay

GDELAY is the key word that invokes the computation of the insertion loss of the resultant two-port.

I.1.11 RCIN—Reflection Coefficient at the Input Port

RCIN is the key word that invokes the computation of the reflection coefficient of the resultant two-port.

I.1.12 RCOUT—Reflection Coefficient at the Output Port

RCOUT is the key word that invokes the computation of the reflection coefficient at the output port of the resultant two-port.

I.1.13 YIN—Input Admittance at the Input Port

YIN is the key word that invokes the computation of the input admittance at the input port of the resultant two-port.

I.1.14 YOUT—Input Admittance at the Output Port

YOUT is the key word that invokes the computation of the input admittance at the output port of the resultant two-port.

I.1.15 ZIN—Input Impedance at the Input Port

ZIN is the key word that invokes the computation of the input impedance at the input port of the resultant two-port.

I.1.16 ZOUT—Input Impedance at the Output Port

ZOUT is the key word that invokes the computation of the input impedance at the output port of the resultant two-port.

I.1.17 STBCIR—Stability Circles

The command STBCIR invokes the computation of the stability factor K_S for the resultant two-port. The input port and output port of the circuit are ports connected to the signal generator port and to the load port, respectively. K_S is a function of frequency and is computed in the frequency band defined in the FREQ block of input data.

The command STBCIR also invokes the computation of the maximum power gain and the related signal generator port reflection coefficient and load port reflection coefficient. These computations take place only in the case of unconditionally stable networks, that is, if the stability factor K_S of the network is bigger than one. For conditionally stable networks ($K_S < 1$), the STBCIR key word invokes the computation of the input port and the output port stability circles. These circles are given by center coordinates (magnitude and phase) and radius lenghts.

I.1.18 PGCIR—Constant Power Gain Circles

For the case $K_S < 1$, JADMIC2 outputs the power gain circle data. This output data is invoked by the command PGCIR.

The x dB circle is requested with the command PGCIR x DB.

Example

PGCIR 6.5 DB

I.1.19 APGCIR—Constant Available Power Gain Circles

For the case $K_S < 1$, JADMIC2 outputs the available power gain circle data. This output data is invoked by the command APGCIR.

The x dB circle is requested with the command APGCIR x DB.

Example

APGCIR 7.5 DB

I.1.20 NFCIR—Constant Noise Figure Circles

The command NFCIR invokes computation of the constant noise figure circle. The x dB circle is requested with the command NFCIR x DB.

Example

NFCIR 2.5 DB

APPENDIX J

Specifying Data Common to All Network Elements—DATA Block

The data block that starts with the mandatory key word DATA is used to define parameter values common to the entire analyzed network.

J.1 ZR STATEMENT—DEFINING REFERENCE IMPEDANCE

The default value for the reference impedance *ZR* common to the entire network equals 50 Ω. It is assumed in JADMIC2 that the same reference impedance is common to all network element ports. If we want to use a reference impedance Z_R that is different than 50 Ω for the network, then the ZR statement is used.

Format

```
ZR = x
```

Example

```
DATA
  ZR = 70 OHM
END
```

Note

If the ZR statement does not appear in the DATA block, the default 50-Ω value for ZR will be used by JADMIC2.

APPENDIX K

Physical Constants Used in JADMIC2

$$\Pi \cong 3.141593$$

Electric constant of vacuum:

$$\varepsilon_o = (4\,\Pi \cdot 10^9)^{-1}\ \text{F/m} \cong 8.854 \cdot 10^{-12}\ \text{F/m}$$

Magnetic constant of vacuum:

$$\mu_o = 4\,\Pi \cdot 10^{-7}\ \text{H/m} \cong 1.257 \cdot 10^{-6}\ \text{H/m}$$

Light velocity in vacuum:

$$c \cong 2.997925 \cdot 10^{11}\ \text{mm/s}$$

Resistivity of gold:

$$\rho_{Ag} = 2.44\ \mu\Omega\ \text{cm}$$

Resistivity of copper:

$$\rho_{Cu} \cong 1.72\ \mu\Omega\ \text{cm}$$

Surface skin resistance (in ohms per square):

$$R_s = \sqrt{\Pi f \mu \rho} \cong 19.869178 \cdot 10^{-4}\ \sqrt{\rho[\Omega cm] \cdot f[\text{Hz}]}$$

Surface skin resistance for copper:

$$R_{s\ Cu} \cong 8.26 \cdot 10^{-3} \sqrt{f[Hz]}$$

Skin depth:

$$\delta = \sqrt{\rho / (\Pi \mu f)}$$

$$\frac{2\,\Pi\, l}{\lambda} = \frac{2\,\Pi}{c} \sqrt{\varepsilon_r} f\, l \cong 2.0958449. \cdot 10^{-11} \sqrt{\varepsilon_r} f\, l$$

About the Author

Janusz A. Dobrowolski was born in Rzeszow, Poland, in 1942. He received M.Sc., Ph.D., and Habilitated Doctor degrees in electronic engineering from the Warsaw University of Technology, Warsaw, Poland, in 1965, 1973, and 1979, respectively.

Since 1965, he has been with the Institute of Electronics Fundamentals, at the Faculty of Electronics and Information Technology, Warsaw University of Technology, Warsaw, Poland. Since 1989 he has held the positions of a Professor and the Director of the Institute.

During the academic years 1974–1975 and 1983–1984, he was a Research Associate in the Department of Electrical Engineering of the University of Manitoba, Winnipeg, Manitoba, Canada, where he worked on the design and development of antennas and microwave components for satellite communication systems.

He has written more than 60 papers and four books. His book *Introduction to Computer Methods for Microwave Circuit Analysis and Design* was published by Artech House in 1991.

His current research interests are in the areas of computer-aided design of microwave circuits and microwave measurement techniques.

He is a Senior Member of the Institute of Electrical and Electronics Engineers, U.S.

Index

The Artech House Microwave Library

Acoustic Charge Transport: Device Technology and Applications, R. Miller, C. Nothnick, and D. Bailey

Advanced Automated Smith Chart Software and User's Manual, Version 2.0, Leonard M. Schwab

Algorithms for Computer-Aided Design of Linear Microwave Circuits, Stanislaw Rosloniec

Analysis, Design, and Applications of Fin Lines, Bharathi Bhat and Shiban K. Koul

Analysis Methods for Electromagnetic Wave Problems, Eikichi Yamashita, editor

Automated Smith Chart Software and User's Manual, Leonard M. Schwab

C/NL2 for Windows: Linear and Nonlinear Microwave Circuit Analysis and Optimization, Software and User's Manual, Stephen A. Maas and Arthur Nichols

Capacitance, Inductance, and Crosstalk Analysis, Charles S. Walker

Computer-Aided Analysis, Modeling, and Design of Microwave Networks: The Wave Approach, Janusz A. Dobrowolski

Design of Impedance-Matching Networks for RF and Microwave Amplifiers, Pieter L. D. Abrie

Dielectric Materials and Applications, Arthur von Hippel, editor

Dielectrics and Waves, Arthur von Hippel

Digital Microwave Receivers, James B. Tsui

Electric Filters, Martin Hasler and Jacques Neirynck

Electrical and Thermal Characterization of MESFETs, HEMTs, and HBTs, Robert Anholt

E-Plane Integrated Circuits, P. Bhartia and P. Pramanick, editors

Feedback Maximization, Boris J. Lurie

Filters with Helical and Folded Helical Resonators, Peter Vizmuller

Frequency Synthesizer Design Toolkit Software and User's Manual, Version 1.0, James A. Crawford

Fundamentals of Distributed Amplification, Thomas T. Y. Wong

GaAs MESFET Circuit Design, Robert A. Soares, editor

GASMAP: Gallium Arsenide Model Analysis Program, J. Michael Golio *et al.*

Handbook of Microwave Integrated Circuits, Reinmut K. Hoffmann

Handbook for the Mechanical Tolerancing of Waveguide Components, W. B. W. Alison

HELENA: HEMT Electrical Properties and Noise Analysis Software and User's Manual, Henri Happy and Alain Cappy

HEMTs and HBTs: Devices, Fabrication, and Circuits, Fazal Ali, Aditya Gupta, and Inder Bahl, editors

High-Power Microwave Sources, Victor Granatstein and Igor Alexeff, edtiors

High-Power GaAs FET Amplifiers, John Walker, editor

High-Power Microwaves, James Benford and John Swegle

Introduction to Microwaves, Fred E. Gardiol

Introduction to Computer Methods for Microwave Circuit Analysis and Design, Janusz A. Dobrowolski

Introduction to the Uniform Geometrical Theory of Diffraction, D. A. McNamara, C. W. I. Pistorius and J. A. G. Malherbe

LINPAR for Windows: Matrix Parameters for Multiconductor Transmission Lines, Software and User's Manual, Antonije Djordjević, Miodrag B. Bazdar, Tapan K. Sarkar, Roger F. Harrington

LOSLIN: Lossy Line Calculation Software and User's Manual, Fred E. Gardiol

Lossy Transmission Lines, Fred E. Gardiol

Low-Angle Microwave Propagation: Physics and Modeling, Adolf Giger

Low Phase Noise Microwave Oscillator Design, Robert G. Rogers

MATCHNET: Microwave Matching Networks Synthesis, Stephen V. Sussman-Fort

Matrix Parameters for Multiconductor Transmission Lines: Software and User's Manual, A. R. Djordjevic *et al.*

Microelectronic Reliability, Volume I: Reliability, Test, and Diagnostics, Edward B. Hakim, editor

Microelectronic Reliability, Volume II: Integrity Assessment and Assurance, Emiliano Pollino, editor

Microwave and Millimeter-Wave Diode Frequency Multipliers, Marek T. Faber, Jerzy Chamiec, Miroslaw E. Adamski

Microwave and RF Circuits: Analysis, Synthesis, and Design, Max Medley

Microwave and RF Component and Subsystem Manufacturing Technology, Heriot-Watt University

Microwave Circulator Design, Douglas K. Linkhart

Microwave Engineers' Handbook, 2 Volumes, Theodore Saad, editor

Microwave Materials and Fabrication Techniques, Second Edition, Thomas S. Laverghetta

Microwave MESFETs and HEMTs, J. Michael Golio *et al.*

Microwave and Millimeter Wave Heterostructure Transistors and Applicatons, F. Ali, editor

Microwave and Millimeter Wave Phase Shifters, Volume I: Dielectric and Ferrite Phase Shifters, S. Koul and B. Bhat

Microwave and Millimeter Wave Phase Shifters, Volume II: Semiconductor and Delay Line Phase Shifters, S. Koul and B. Bhat

Microwave Mixers, Second Edition, Stephen Maas

Microwave Transmission Design Data, Theodore Moreno

Microwave Transition Design, Jamal S. Izadian and Shahin M. Izadian

Microwave Transmission Line Couplers, J. A. G. Malherbe

Microwave Tubes, A. S. Gilmour, Jr.

Microwaves: Industrial, Scientific, and Medical Applications, J. Thuery

Microwaves Made Simple: Principles and Applicatons, Stephen W. Cheung, Frederick H. Levien *et al.*

MMIC Design: GaAs FETs and HEMTs, Peter H. Ladbrooke

Modern GaAs Processing Techniques, Ralph Williams

Modern Microwave Measurements and Techniques, Thomas S. Laverghetta

Monolithic Microwave Integrated Circuits: Technology and Design, Ravender Goyal *et al*

Nonuniform Line Microstrip Directional Couplers, Sener Uysal

PC Filter: Electronic Filter Design Software and User's Guide, Michael G. Ellis, Sr.

PLL: Linear Phase-Locked Loop Control Systems Analysis Software and User's Manual, Eric L. Unruh

RF Design Guide: Systems, Circuits, and Equations, Peter Vizmuller

Scattering Parameters of Microwave Networks with Multiconductor Transmission Lines: Software & User's Manual, A. R. Djordjevic *et al.*

Solid-State Microwave Power Oscillator Design, Eric Holzman and Ralston Robertson

Terrestrial Digital Microwave Communications, Ferdo Ivanek *et al.*

Time-Domain Response of Multiconductor Transmission Lines: Software and User's Manual, A. R. Djordjevic *et al.*

Transmission Line Design Handbook, Brian C. Waddell

TRAVIS Pro: Transmission Line Visualization Software and User's Manual, Professional Version, Robert G. Kaires and Barton T. Hickman

TRAVIS Student: Transmission Line Visualization Software and User's Manual, Student Version, Robert G. Kaires and Barton T. Hickman

Yield and Reliability in Microwave Circuit and System Design, Michael Meehan and John Purviance

For further information on these and other Artech House titles, contact:

Artech House
685 Canton Street
Norwood, MA 02062
617-769-9750
Fax: 617-769-6334
Telex: 951-659
e-mail: artech@world.std.com

Artech House
Portland House, Stag Place
London SW1E 5XA England
+44 (0) 171-973-8077
Fax: +44 (0) 171-630-0166
Telex: 951-659
e-mail: bookco@artech.demon.co.uk